Numerical Solutions of the N-body Problem

Mathematics and Its Applications (*East European Series*)

Managing Editor:

M. HAZEWINKEL

Centre for Mathematics and Computer Science, Amsterdam, The Netherlands

Editorial Board:

A. BIALYNICKI-BIRULA, *Institute of Mathematics PKIN, Warsaw, Poland*
J. KURZWEIL, *Mathematics Institute, Academy of Sciences, Prague, Czechoslovakia*
L. LEINDLER, *Bolyai Institute, Szeged, Hungary*
L. LOVÁSZ, *Bolyai Institute, Szeged, Hungary*
D. S. MITRINOVIĆ, *University of Belgrade, Yugoslavia*
S. ROLEWICZ, *Polish Academy of Sciences, Warsaw, Poland*
BL. H. SENDOV, *Bulgarian Academy of Sciences, Sofia, Bulgaria*
I. T. TODOROV, *Academy of Sciences, Sofia, Bulgaria*
H. TRIEBEL, *Universität Jena, D.D.R.*

Andrzej Marciniak

Institute of Mathematics,
A. Mickiewicz University, Poznan, Poland

Numerical Solutions of the N-body Problem

D. Reidel Publishing Company

A MEMBER OF THE KLUWER ACADEMIC PUBLISHERS GROUP

Dordrecht / Boston / Lancaster

Library of Congress Cataloging in Publication Data

CIP

Marciniak, Andrzej, 1953–
Numerical solutions of the N-body problem.

(Mathematics and its applications. East European series)
Includes index.
1. Many-body problem–Numerical solutions. 2. Initial value problems–Numerical solutions. I. Title. II. Series.
QA378.M37 1985 530.1'44 85–10821
ISBN 90–277–2058–4

Published by D. Reidel Publishing Company
P.O. Box 17, 3300 AA Dordrecht, Holland

Sold and distributed in the U.S.A. and Canada
by Kluwer Academic Publishers,
190 Old Derby Street, Hingham, MA 02043, U.S.A.

In all other countries, sold and distributed
by Kluwer Academic Publishers Group,
P.O. Box 322, 3300 AH Dordrecht, Holland

Printed in The Netherlands

To My Wife

Contents

SERIES EDITOR'S PREFACE

Approach your problem from the right end and begin with the answers. Then one day, perhaps you will find the final question.

The Hermit Clad in Crane Feathers in R. van Gulik's The Chinese Maze Murders.

It isn't that they can't see the solution. It is that they can't see the problem.

G.K. Chesterton. The Scandal of Father Brown The Point of a Pin.

Growing specialization and diversification have brought a host of monographs and textbooks on increasingly specialized topics. However, the "tree" of knowledge of mathematics and related fields does not grow only by putting forth new branches. It also happens, quite often in fact, that branches which were thought to be completely disparate are suddenly seen to be related.

Further, the kind and level of sophistication of mathematics applied in various sciences has changed drastically in recent years: measure theory is used (non-trivially) in regional and theoretical economics; algebraic geometry interacts with physics; the Minkowsky lemma, coding theory and the structure of water meet one another in packing and covering theory; quantum fields, crystal defects and mathematical programming profit from homotopy theory; Lie algebras are relevant to filtering; and prediction and electrical engineering can use Stein spaces. And in addition to this there are such new emerging subdisciplines as "experimental mathematics", "CFD", "completely integrable systems", "chaos, synergetics and large-scale order", which are almost impossible to fit into the existing classification schemes. They draw upon widely different sections of mathematics. This programme, Mathematics and Its Applications, is devoted to new emerging (sub)disciplines and to such (new) interrelations as exempla gratia:

- a central concept which plays an important role in several different mathematical and/or scientific specialized areas;
- new applications of the results and ideas from one area of scientific endeavour into another;
- influences which the results, problems and concepts of one field of enquiry have and have had on the development of another.

The Mathematics and Its Applications programme tries to make available a careful selection of books which fit the philosophy outlined above. With such books, which are stimulating rather than definitive, intriguing rather than encyclopaedic, we hope to contribute something towards better communication among the practitioners in diversified fields.

The N-body problem is an old one and, in spite of the efforts of the best mathematicians of all time, unsolved for $N>3$, and only really solved for $N=2$ (where of course the solution is easy and almost 400 years old). The case $N=3$ is intermediate and would require at least a full length book to describe all that is known. We know, thanks to the work of Siegel, Kolmogorov, Arnol'd, Moser, and others, that the dynamics are exceedingly complicated in terms of their dependence on initial conditions.
Numerical simulation has wrought revolutions in several parts of mathematics. Whether it can also perform similar services for this old, exceedingly intractable, problem is something I would not dare try to predict. But certainly insight can be generated this way both for veterans and newcomers to the field. This requires the appropriate numerics which takes full account of the special properties of this particular problem. This is exactly what this book provides.

The unreasonable effectiveness of mathematics in science ...

Eugene Wigner

Well, if you know of a better 'ole, go to it.

Bruce Bairnsfather

What is now proved was once only imagined.

William Blake.

As long as algebra and geometry proceeded along separate paths, their advance was slow and their applications limited.
But when these sciences joined company they drew from each other fresh vitality and thenceforward marched on at a rapid pace towards perfection.

Joseph Louis Lagrange.

Bussum, March 1985

Michiel Hazewinkel

Preface

This book is basically designed for university students majoring in numerical methods and their applications. It contains a review of conventional numerical methods used for solving the initial value problem, an analysis of their applications to different variants of the N-body problem and analysis of some special methods of solving this problem. In addition, it presents more than twenty computer algorithms which can immediately be programmed in any programming language. For this reason, this book may also be of interest to professionals involved in research in the N-body problem.

I wrote this book during my stay at the University of Florida in Gainesville on an exchange program with A.Mickiewicz University in Poznań, Poland. I express my sincere gratitude to Prof.A.R.Bednarek, chairman of the Department of Mathematics, for creating conditions in which I was able to write the book. I also address special thanks to Prof.J.Albrycht of Poznań's Technical University without whose inspiration, encouragement and advice this book would not have been written.

Gainesville, May, 1984 Andrzej Marciniak

Introduction

The N-body problem, i.e. the problem of determining the motion of N material points, has been known since Newton formulated his law of gravity. The greatest mathematicians have tried to solve it, but so far an analytical solution of the N-body problem for $N > 3$ has not been found.

The development of digital computers has contributed to our ability to solve the N-body problem numerically. The solution obtained, however, no matter what numerical method was used, requires a determination of accuracy, as well as a study of stability and convergence. It is also important to choose the most effective method. The purpose of this book is - among other things - an analysis of the application of conventional numerical methods to solve the different variants of the N-body problem. Moreover, some special numerical methods which conserve the constants of motion (e.g. linear and angular momenta, total energy, Jacobi's constant) are presented. For the majority of the conventional and all the special numerical methods of solving different variants of the N-body problem, the algorithms presented in this book can immediately be programmed in any programming language.

Chapter 1 aims at presenting to the reader the elements of numerical analysis and the conventional numerical methods used for solving the initial value problem. One-step methods with automatic step choice, extrapolation methods, and especially the commonly used rational extrapolation method of Gragg-Bulirsch-Stoer deserve special attention.

In Chapter 2, a short introduction to the N-body problem and an example of a solution of it for N=2 are followed by an analysis of the application of conventional numerical methods. Next, the discrete mechanics of Greenspan, i.e. the numerical method which conserves the constants of motion and whose formulas are invariant with respect to some transformations of the frame of reference, is presented. Furthermore, Chapter 2 contains a description of the discrete mechanics of arbitrary order and a modification of polynomial extrapolation which, when used for an arbitrary conventional methods, ensures a conservation of the total energy by the obtained numerical solution.

Chapter 3 is devoted to the application of conventional numerical methods and discrete mechanics formulas for the solution of the problem of relative motion of N bodies.

The equations of motion in rotating frames of reference are dealt with in Chapter 4. An analysis of applications of conventional numerical methods in Sect.4.3 is followed by a presentation of discrete dynamical equations conserving Jacobi's constant of motion. Furthermore, Chapter 4 examines in greater detail the application of these equations to the study of motion of a body with an infinitely small mass nearby the equilibrium points and to the study of moon motion.

In the Appendix, for all nonconventional, described in this book, numerical methods of solving different variants of the N-body problem, the procedures in the PL/I language for computers are given.

It is obvious that the problems relating to an application of the conventional and special numerical methods for solving the equations of motion of material points which are described in this book do not exhaust the topic of numerical solutions of the N-body problem. Their only purpose is to present to the reader some questions and to give the ways of solution of them. If, however, the reader develops a critical attitude toward the ideas presented here, modifies the given algorithms or creates his own, more effective algorithms, the author will know that the aim of this book has been achieved.

Chapter 1

Conventional Numerical Methods for Solving the Initial Value Problem

1.1. THE INITIAL VALUE PROBLEM

Let us consider the following system of ordinary first--order differential equations

$$y' = f(x,y), \tag{1.1.1}$$

where $y, f(x,y) \in R^m$, $x \in [a,b]$. The function $z=z(x)$ is called a solution of (1.1.1) if z is defined and differentiable for $x \in [a,b]$ and

$$z'(x) = f[x,z(x)].$$

In general, there exists a set of solutions containing all solutions without exception. This set is called a general solution and depends on an arbitrary constant. For example, for $m=1$ the equation $y'=y$ has the general solution of the form $z(x)=Ce^x$, where C denotes an arbitrary real number. In order to secure the uniqueness of solution we impose an additional condition on the function y. It is usually the initial condition

$$y(a) = z^o; \quad z^o \in R^m \tag{1.1.2}$$

or the boundary one

$$g[y(a),y(b)] = 0,$$

where g denotes the given function of $2m$ variables.

In the N-body problem the equations of motion have the form

$$y'' = f(y),$$

where $y, f(y) \in R^{3N}$ and compose a system of $3N$ second-order differential equations. However, substituting $y'=\hat{y}$, we can always reduce (1.1.3) to (1.1.1). Since in the N-body problem an initial condition is given, then in a later passage of this book we will consider only the problem (1.1.1)-(1.1.2). A problem of obtention of the function z, which is a solution of (1.1.1) and fulfils the initial condition (1.1.2), is called the initial value problem.

From the theory of ordinary differential equations it is known that if the function f fulfils some simple conditions of regularity, then there exists one and only one solution of the problem (1.1.1)-(1.1.2). Namely, we have

Theorem 1.1. Let f be defined and continuous for $(x,y)\in\{(t,w): t\in[a,b],\ w\in R^m\}$. Moreover, let us assume that f fulfils the Lipschitz condition with respect to the second variable, i.e. there exists L=const that for any $x\in[a,b]$ and $y_1,y_2\in R^m$ we have

$$\| f(x,y_1)-f(x,y_2) \| \le L\| y_1-y_2 \|, \tag{1.1.4}$$

where $\| . \|$ denotes a norm in R^m.

Then for each $x^o\in[a,b]$ and $z^o\in R^m$ there exists one and only one function $z=z(x)$ such that:

(i) $z(x)$ is continuous and continuously differentiable for $x\in[a,b]$;

(ii) $z'(x)=f[x,z(x)]$ for $x\in[a,b]$;

(iii) $z(x^o)=z^o$.

The proof of the above theorem can be found in every handbook for differential equations (see e.g. [16], [71], [72], [82]) and therefore we omit one.

The Lipschitz condition holds especially if in the set $\{(t,w): t\in[a,b],\ w\in R^m\}$ there exist continuous and bounded partial derivatives $\partial f_i/\partial y_j$ $(i,j=1(1)m)$. If the partial derivatives are unbounded in this set, then the initial value problem (1.1.1)-(1.1.2) has a solution only in some neighbourhood of x^o (instead of on the whole interval [a,b]). Further on we will always assume that there exists a unique solution of (1.1.1)-(1.1.2) on the whole interval [a,b].

For the later considerations, relating especially to analysis of numerical methods for (1.1.1)-(1.1.2), it is convenient to write the initial value problem in another way ([86]). Without loss of generality let us assume that [a,b]=[0,1]. If $x\in[a,b]$, then substituting $\bar{x}=(x-a)/(b-a)$, $a\neq b$, we can always carry the interval of definition of the independent variable to [0,1].

Let $E=\times_m C^{(1)}[0,1]$ ($\times_m$ denotes a m-tuply product) and $E^o=\times_m R \times \times_m C^{(1)}[0,1]$. In the spaces E and E^o we introduce the norms as follows

$$
\begin{aligned}
\| y \|_E &= \max_{x\in[0,1]} \Big(\sum_{i=1}^{m} |y_i(x)| \Big), \\
\| \bar{d} \|_{E^o} &= \left\| \begin{bmatrix} d^o \\ d \end{bmatrix} \right\|_{E^o} = \sum_{i=1}^{m} |d_i^o| \\
&\quad + \max_{x\in[0,1]} \Big(\sum_{i=1}^{m} |d_i(x)| \Big).
\end{aligned}
\tag{1.1.5}
$$

Let $F: E\to E^o$. Then the initial value problem (1.1.1)-(1.1.2) can be rewritten in the form

$$
Fy = \begin{bmatrix} y(0) - z^o \\ y' - f(x,y) \end{bmatrix} \in E^o \text{ for } y\in E. \tag{1.1.6}
$$

By an exact solution of (1.1.6) we understand such element $z\in E$ that $Fz=0$. The initial value problem (1.1.6) will be also briefly denoted by $B=\{E,E^o,F\}$.

1.2. DISCRETIZATION METHODS

In this section, basing oneself on [86], we adduce some notions relating to the discretization of the initial value problem (1.1.6) and numerical analysis of discrete methods. At first, let us introduce the following notion:

Definition 1.1. The method M applied to $B=\{E,E^o,F\}$ is called a discretization method, if $M=\{E_n,E_n^o,\Delta_n,\Delta_n^o,\phi_n\}_{n\in N'}$, where $N'\subset N$, $E_n=(G_n\to R^m)$, $E_n^o=(G_n\to R^m)$, $G_n=\{t_\nu:\ t_o=0,\ t_{\nu-1}<t_\nu \text{ for } \nu=1(1)n\}$ and where $\Delta_n: E\to E_n$, $\Delta_n^o: E^o\to E_n^o$ denote linear mappings such that

$\lim_{n\to\infty} \| \Delta_n y \|_{E_n} = \| y \|_E$ for each fixed $y\in E$,

$\lim_{n\to\infty} \| \Delta_n^o \bar{d} \|_{E_n^o} = \| \bar{d} \|_{E^o}$ for each fixed $\bar{d}\in E^o$, $\phi_n: (E\to E^o)\to(E_n\to E_n^o)$, where F belongs to a domain of every ϕ_n.

The notation $(G_n \to R^m)$ denotes a space whose elements are to be the functions defined on the net G_n with the values in R^m. In the above definition the net G_n is arbitrary. In practice we most often deal with the uniform net, i.e. $G_n=\{\nu/n: \nu=0(1)n\}$. In the analogy of (1.1.4) we introduce the norms in E_n and E_n^o respectively as follows (for a uniform net)

$$\| \eta \|_{E_n} = \max_{\nu=0(1)n} \left(\sum_{i=1}^{m} |\eta_i(\tfrac{\nu}{n})| \right),$$
$$\| \delta \|_{E_n^o} = \sum_{i=1}^{m} |\delta_i(0)| + \max_{\nu=1(0)n} \left(\sum_{i=1}^{m} |\delta_i(\tfrac{\nu}{n})| \right). \tag{1.2.1}$$

<u>Definition 1.2.</u> If $D=\{E_n, E_n^o, F_n\}$, where $F_n: E_n \to E_n^o$, then D is said to be a discrete problem. The sequence $\{\zeta_n\}_{n\varepsilon N'}$, where $\zeta_n \varepsilon E_n$, is called a solution of D if $F_n\zeta_n=0$ for $n\varepsilon N'$. The discrete problem D is called a discrete initial value problem for the initial value problem B if D is obtained by a discretization method M applied to B and $F_n=\phi_n(F)$. In this case D will be also denoted by M(B).

The relationships between mappings occuring in Defs.1.1 and 1.2 are illustrated in Fig.1.

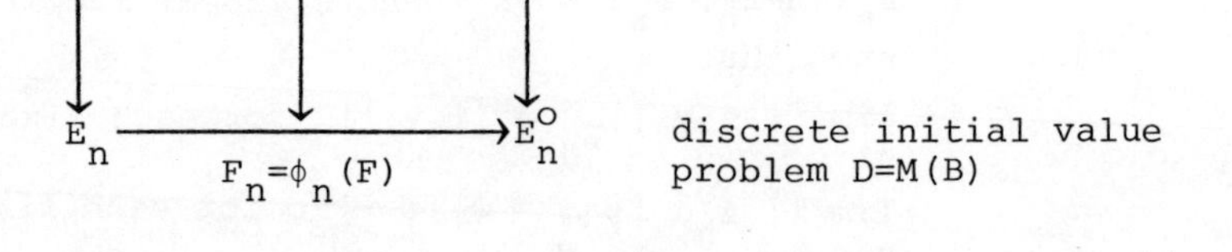

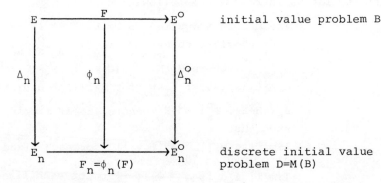

Fig.1. Relationships between elements of discretization method

The last definition states that by applying a discretization method to the initial value problem (1.1.6), we obtain a discrete initial value problem. This discrete problem should approximate the initial value problem in the meaning below.

Definition 1.3. We say that the method M is consistent with B on the element $y \in E$ if y belongs to the domains of F and $\phi_n(F)\Delta_n$ $(n \in N')$ and if

$$\lim_{n\to\infty} || \phi_n(F)\Delta_n y - \Delta_n^o Fy ||_{E_n^o} = 0.$$

If M is consistent with B on all $y \in E$, then M is called consistent with B. M and D=M(B) are called consistent with B with order p on y (or shortly: of order p) if

$$|| \phi_n(F)\Delta_n y - \Delta_n^o Fy ||_{E_n} = O(n^{-p}) \text{ as } n\to\infty.$$

A solution of the discrete initial value problem should be convergent to the solution of the adequate initial value problem of the form (1.1.6).

Definition 1.4. If the discretization method M is applied to the initial value problem B with the exact solution z and the discrete initial value problem M(B) has the unique solution $\{\zeta_n\}_{n\in N'}$, then the sequence $\{\varepsilon_n\}_{n\in N'}$, where

$$\varepsilon_n = \zeta_n - \Delta_n z \in E_n; \quad n \in N', \tag{1.2.2}$$

is called a total discretization error of M and M(B) on B. If

$$\lim_{n\to\infty} || \varepsilon_n ||_{E_n} = 0,$$

then we say that the method M and problem M(B) are convergent on B. M and M(B) are said to be convergent of order p on B if

$$|| \varepsilon_n ||_{E_n} = O(n^{-p}) \text{ as } n\to\infty.$$

Let us note that a discretization method consistent with the initial value problem may be convergent, but it does not have to be. The examples are presented, among others, in [54] and [86].

It is obvious that any numerical realization of the method M in a floating-point arithmetic on computers yields only an approximation of exact solution with beforehand specified and finite precision. So, the convergence is only a theoretical property of discretization methods.

Another basic requirement for the discrete methods is a stability of them. From (1.2.2) it follows that a total discretization error is a difference between the solutions of equations $F_n\zeta_n=0$ and $F_n\zeta_n'=\phi_n(F)\Delta_n z$. If we assume that the initial value problem has a unique solution, then the second equation has the solution $\zeta_n'=\Delta_n z$. To decrease the total discretization error ε_n together with $\phi_n(F)\Delta_n z$ it is sufficient to have an estimation of perturbation of the equation $F_n\zeta_n=0$ independent of n. Insensibility to perturbations is just the stability.

Definition 1.5. The discrete problem D is called stable on the sequence $\eta=\{\eta_n\}_{n\in N'}$ $(\eta_n\in E_n)$ if there exist constants S and $r>0$ independent of n such that

$$|| \eta_n^{(1)} - \eta_n^{(2)} ||_{E_n} \leq S|| F_n\eta_n^{(1)} - F_n\eta_n^{(2)} ||_{E_n^o}$$

for all $\eta_n^{(i)}$ $(i=1,2)$ for which

$$|| F_n\eta_n^{(i)} - F_n\eta_n ||_{E_n^o} < r.$$

If $D=M(B)$ is stable on $\{\Delta_n z\}$, then the method M and the discrete initial value problem D are said to be stable on B.

The basic theorem that joins the notions of consistency, stability and convergence for discretization methods is the following one:

Theorem 1.2. Let for the initial value problem $B=\{E,E^o,F\}$ with the exact solution z the discretization method $M=\{E_n,E_n^o,\Delta_n,\Delta_n^o,\phi_n\}_{n\in N'}$, applied to B fulfils the conditions:

(i) the mappings $F_n=\phi_n(F)$: $E_n\to E_n^o$ are determined and continuous in the domain $B_R(\Delta_n z)$ $=\{\eta_n\in E_n: || \eta_n - \Delta_n z || < R\}$, where R is independent of n;

(ii) M is consistent with B on z;

(iii) M is stable on B.

Then the discrete initial value problem M(B) has the unique solution $\{\zeta_n\}$, $\zeta_n\in E_n$, for all sufficiently small $n \in N' \subset N$ and the discretization method M is convergent on B. Incidentally, if M is consistent with B with order p, then M is convergent on B with the same order.

The proof of this theorem can be found in [86].

Taking into account the well-known Euler method with m=1, i.e. applied to a scalar differential equation, we now illustrate the notions mentioned above.

Let $E=C^{(1)}[0,1]$ and $E^o=R\times C^{(1)}[0,1]$. We have $E_n=(G_n\to R)$ and $E_n^o=(G_n\to R)$. Let us assume that the net G_n is uniform. The linear mappings $\Delta_n:E\to E_n$ and $\Delta_n^o:E^o\to E_n^o$ we define as follows

$$(\Delta_n y)(\tfrac{\nu}{n}) = y(\tfrac{\nu}{n}) \text{ for } y\in E,$$

$$(\Delta_n^o d)(\tfrac{\nu}{n}) = \begin{cases} d^o, & \nu=0, \\ d(\frac{\nu-1}{n}), & \nu=1(1)n, \end{cases} \text{ for } d=\begin{bmatrix} d^o \\ d(t)\end{bmatrix}\in E^o$$

and the mapping ϕ_n by

$$[\phi_n(F)\eta](\tfrac{\nu}{n}) = \begin{cases} \eta(0)-z^o, & \nu=0, \\ \dfrac{\eta(\frac{\nu}{n})-\eta(\frac{\nu-1}{n})}{1/n} - f[\frac{\nu-1}{n},\eta(\frac{\nu-1}{n})], & \nu=1(1)n. \end{cases} \tag{1.2.3}$$

The Euler method is consistent with the initial value problem (1.1.6), because

$$[\phi_n(F)\Delta_n y-\Delta_n^o Fy](\tfrac{\nu}{n}) = \begin{cases} (y(0)-z^o)-(y(0)-z^o), & \nu=0, \\ \dfrac{y(\frac{\nu}{n})-y(\frac{\nu-1}{n})}{1/n} - f[\frac{\nu-1}{n},y(\frac{\nu-1}{n})] \\ \quad - y'(\frac{\nu-1}{n})-f[\frac{\nu-1}{n},y(\frac{\nu-1}{n})], & \nu=1(1)n \end{cases}$$

$$= \begin{cases} 0, & \nu=0, \\ y'(\xi_\nu)-y'(\frac{\nu-1}{n}), & \nu=1(1)n, \end{cases}$$

where $\xi_\nu\in(\frac{\nu-1}{n},\frac{\nu}{n})$. Since

$$\max_{\nu=1(1)n} \left| y'(\xi_\nu)-y'(\tfrac{\nu-1}{n}) \right| \to 0 \text{ as } n\to\infty,$$

then

$$\lim_{n\to\infty} \| \phi_n(F)\Delta_n y - \Delta_n^o Fy \|_{E_n^o} = 0$$

and this means that the method (1.2.3) is consistent.

Now, let us assume that the function f in the initial value problem (1.1.6) fulfils the Lipschitz condition with respect to the second variable. Thus

$$\begin{aligned}|y'(\xi_\nu)-y'(\tfrac{\nu-1}{n})| &= |f[\xi_\nu,y(\xi_\nu)]-f[\tfrac{\nu-1}{n},y(\tfrac{\nu-1}{n})]|\\ &\le L|y(\xi_\nu)-y(\tfrac{\nu-1}{n})|\\ &= L|y(\xi_\nu)-y(\xi_\nu)+\tfrac{1}{n}y'(\theta_\nu)| = \tfrac{L}{n}|y'(\theta_\nu)|,\end{aligned}$$

where $\theta_\nu \epsilon (\frac{\nu-1}{n},\xi_\nu)$. Since $y\epsilon C^{(1)}[0,1]$, then there exists $M>0$ such that for each $x\epsilon[0,1]$ we have $|y'(x)|\le M$. Hence we obtain

$$||\ \phi_n(F)\Delta_n y - \Delta_n^o Fy\ ||_{E_n^o} \le \frac{LM}{n},$$

and so $||\phi_n(F)\Delta_n y-\Delta_n^o Fy||_{E_n^o}=O(n^{-1})$, i.e. Euler's method is consistent with B with order 1. Let us note that if ϕ_n is defined, instead of (1.2.3), by the formula

$$[\phi_n(F)\eta](\tfrac{\nu}{n}) = \begin{cases}\eta(0)-z^o, & \nu=0,\\ \eta(\frac{\nu}{n})-\eta(\frac{\nu-1}{n})-\frac{1}{n}f[\frac{\nu-1}{n},\eta(\frac{\nu-1}{n})], & \nu=1(1)n,\end{cases}$$

then this method will not be consistent with the initial value problem (1.1.6) in the meaning of Def.1.3.

Now, let in the problem (1.1.6) $f(x,y)=y$. Then the equation (1.1.6) has the solution $z=z^o e^x$ and hence $(\Delta_n z)(\frac{\nu}{n})=z^o e^{\nu/n}$, whereas the discrete initial value problem (1.2.3) has the solution $\zeta_n(\frac{\nu}{n})=z^o(1+\frac{1}{n})^\nu$. So, we have

$$\varepsilon_n(\tfrac{\nu}{n}) = z^o[(1+\tfrac{1}{n})^\nu-e^{\nu/n}], \text{ for } \nu=0(1)n$$

and

$$\begin{aligned}||\ \varepsilon_n\ ||_{E_n} &= \max_{\nu=0(1)n} |\varepsilon_n(\tfrac{\nu}{n})|\\ &= |z^o| \max_{\nu=0(1)n} |(1+\tfrac{1}{n})^\nu-e^{\nu/n}|\\ &= |z^o||(1+\tfrac{1}{n})^n-e|\\ &= |z^o||1+n\tfrac{1}{n}+\frac{n(n-1)}{2!}\frac{1}{n^2}+\frac{n(n-1)(n-2)}{3!}\frac{1}{n^3}\\ &\qquad +\ldots -e|\\ &= |z^o||1+\tfrac{1}{1!}+\tfrac{1}{2!}+\ldots-\tfrac{1}{2n}(1+\tfrac{1}{1!}+\tfrac{1}{2!}\\ &\qquad +\ldots)+O(n^{-2})-e|\end{aligned}$$

$$= |z^o||e- \frac{1}{2n} e+O(n^{-2})-e| = \frac{1}{2n}|z^o|e+O(n^{-2}).$$

This means that the method (1.2.3) is convergent with the first order.

Let us assume once more that the function f in (1.1.6) fulfils Lipshitz's condition with respect to the second variable, i.e. for arbitrary $y_i \in R$ (i=1,2)

$$|f(x,y_1)-f(x,y_2)| \le L|y_1-y_2|. \tag{1.2.4}$$

To show stability of the method (1.2.3), let us take arbitrary $\eta_n^{(i)} \in E_n$ (i=1,2). From (1.2.3) we have

$$\eta_n^{(1)}(\tfrac{\nu}{n})-\eta_n^{(2)}(\tfrac{\nu}{n}) = \begin{cases} F_n\eta_n^{(1)}(0)-F_n\eta_n^{(2)}(0), \quad \nu=0, \\ \eta_n^{(1)}(\frac{\nu-1}{n})-\eta_n^{(2)}(\frac{\nu-1}{n}) \\ \quad +\frac{1}{n} f[\frac{\nu-1}{n},\eta_n^{(1)}(\frac{\nu-1}{n})]-f[\frac{\nu-1}{n},\eta_n^{(2)}(\frac{\nu-1}{n})] \\ \quad +\frac{1}{n}[F_n\eta_n^{(1)}(\frac{\nu}{n})-F_n\eta_n^{(2)}(\frac{\nu}{n})], \quad \nu=1(1)n. \end{cases}$$

Hence and from (1.2.4) we obtain

$$\begin{aligned} |\eta_n^{(1)}(\tfrac{\nu}{n})-\eta_n^{(2)}(\tfrac{\nu}{n})| &\le |\eta_n^{(1)}(\tfrac{\nu-1}{n})-\eta_n^{(2)}(\tfrac{\nu-1}{n})|(1+\tfrac{L}{n}) \\ &\quad + \tfrac{1}{n}|F_n\eta_n^{(1)}(\tfrac{\nu}{n})-F_n\eta_n^{(2)}(\tfrac{\nu}{n})| \\ &\le |\eta_n^{(1)}(0)-\eta_n^{(2)}(0)|(1+\tfrac{L}{n})^{\nu} \\ &\quad + \tfrac{1}{n}\sum_{\mu=1}^{\nu} |F_n\eta_n^{(1)}(\tfrac{\nu}{n})-F_n\eta_n^{(2)}(\tfrac{\nu}{n})|(1+\tfrac{L}{n})^{\nu-\mu} \\ &= |F_n\eta_n^{(1)}(0)-F_n\eta_n^{(2)}(0)|(1+\tfrac{L}{n})^{\nu} \\ &\quad + \tfrac{1}{n}\sum_{\mu=1}^{\nu} |F_n\eta_n^{(1)}(\tfrac{\mu}{n})-F_n\eta_n^{(2)}(\tfrac{\mu}{n})|(1+\tfrac{L}{n})^{\nu-\mu}. \end{aligned}$$

Therefore,

$$\begin{aligned} ||\eta_n^{(1)}-\eta_n^{(2)}||_{E_n} &= \max_{\nu=0(1)n} |\eta_n^{(1)}(\tfrac{\nu}{n})-\eta_n^{(2)}(\tfrac{\nu}{n})| \\ &\le |F_n\eta_n^{(1)}(0)-F_n\eta_n^{(2)}(0)|(1+\tfrac{L}{n})^{n} \\ &\quad + \tfrac{1}{n}\sum_{\nu=1}^{n} |F_n\eta_n^{(1)}(\tfrac{\nu}{n})-F_n\eta_n^{(2)}(\tfrac{\nu}{n})| \quad (1+\tfrac{L}{n})^{n-\nu} \\ &\le e^{L}[|F_n\eta_n^{(1)}(0)-F_n\eta_n^{(2)}(0)| \\ &\quad + \max_{\nu=1(1)n} |F_n\eta_n^{(1)}(\tfrac{\nu}{n})-F_n\eta_n^{(2)}(\tfrac{\nu}{n})|] \end{aligned}$$

$$= e^L || F_n \eta_n^{(1)} - F_n \eta_n^{(2)} ||_{E_n^o} .$$

Thus, the method (1.2.3) is stable on $\{\eta_n\}$ with $S=e^L$ and $r=\infty$. Since $\{\eta_n\}$ is an arbitrary sequence in the space E_n, then, taking $\eta_n=\Delta_n z$, we get that the discrete initial value problem (1.2.3) is stable on the initial value problem (1.1.6).

Let us note that the proof of consistency and stability for the method (1.2.3) is independent of the form of f. It depends only on the fulfilment of the Lipschitz condition. However, to show a convergence of Euler's method the form of f plays a central role. Since in a lot of problems we do not know an analytical form of the solution of (1.1.6), then we can not use directly the Def.1.4 to prove a convergence of an adequate discrete method. Owing to Th.1.2, we can do it by showing the consistency, stability and determination with continuity of mappings F_n in some domains.

In the practical realization of the initial value problem on a digital computer it is important to determine, using different solution accuracies given beforehand, the following quantities:

(i) total time to integrate over the range;
(ii) overhead, i.e. the total time, excluding derivative evaluations;
(iii) the number of derivative evaluations, including, if necessary, the derivatives of higher orders;
(iv) the number of integration steps (for methods with an automatic step size correction);
(v) the total discretization error.

As we have noted above, the determination of total discretization error is often impossible. If the analytical solution is not known, then we apply the following procedure ([44]): In each point ν/n (if we use the uniform net), where $\nu=1(1)n$, we calculate the quantity

$$l_\nu = \eta(\tfrac{\nu}{n}) - u(\tfrac{\nu}{n},\tfrac{\nu-1}{n}), \tag{1.2.5}$$

where $u(x,\frac{\nu-1}{n})$ denotes a solution of the initial value problem

$$\begin{aligned} u' &= f(x,u); \quad x\epsilon[\tfrac{\nu-1}{n},1]; \\ u(\tfrac{\nu-1}{n},\tfrac{\nu-1}{n}) &= \eta(\tfrac{\nu-1}{n}). \end{aligned} \tag{1.2.6}$$

So, we obtain the value $u(\frac{\nu}{n},\frac{\nu-1}{n})$ if we solve numerically with sufficiently great precision this initial value problem on the interval $[(\nu-1)/n,\nu/n]$. If $\bar{\varepsilon}$ denotes a solution accuracy given beforehand, then the total discretization error we determine from the formula

$$|| \varepsilon_n ||_{E_n} = \max_{\nu=1(1)n} \frac{n||1_\nu||}{\bar{\varepsilon}} . \qquad (1.2.7)$$

The formulas (1.2.5)-(1.2.7) are also correct for an arbitrary net $G_n=\{t_\nu:\ t_o=0,\ t_{\nu-1}<t_\nu,\ \nu=1(1)n\}$, but the quantities $(\nu-1)/n$ and ν/n must be replaced by $t_{\nu-1}$ and t_ν respectively.

The notation of discretization methods and the discrete initial value problem introduced by H.J.Stetter and presented in this chapter is very convenient and useful in a numerical analysis of different methods. Since in a lot of handbooks and monographs devoted to numerical methods for solving the initial value problem, the conventional notation of these methods is used, we will use one under discussion of particular methods. The notation of Stetter will be applied to an analysis of numerical methods for solving the N-body problem in the next chapters.

Discretization methods are presented in a lot of books. Apart from the book of Stetter [86], one ought to make mention of [32],[44],[47],[59],[60],[68],[69] and [80]. Numerical methods for solving the initial value problem and their analysis are also discussed in [4],[21],[52],[54],[76] and [87].

1.3. SURVEY OF NUMERICAL METHODS FOR THE INITIAL VALUE PROBLEM

Our objective in this section is to briefly present the conventional numerical methods for solving the initial value problem (1.1.1)-(1.1.2) and the ways of practical realization of these methods on digital computers. We will not consider in detail a numerical analysis of them and will limit our interest to the citation of some results. Further details the reader can find in cited references.

Taking into account the capacity of this book, it is not advisable to present all well-known numerical methods for solving the initial value problem. Therefore, our attention will be brought only to the methods most popular for solving the problems of celestial mechanics.

1.3.1. ONE-STEP METHODS

Let us assume that the initial value problem (1.1.1)-(1.1.2) has the solution $z=z(x)\epsilon R^m$ differentiable (s+1)-times in the interval [a,b]. Applying the Taylor formula, we have

$$z(x+h) = z(x)+hz'(x)+\frac{h^2}{2!}z''(x)+\ldots+\frac{h^s}{s!}z^{(s)}(x) +R(x,h), \tag{1.3.1}$$

where $R(x,h)=O(h^{s+1})$. From (1.1.1) we obtain

$$\begin{aligned} z'(x) &= f(x,z), \\ z''(x) &= f'(x,z) = \frac{\partial f}{\partial x}(x,z) + \sum_{i=1}^{m} \frac{\partial f}{\partial z_i}(x,z)f_i(x,z), \\ z'''(x) &= f''(x,z) = \frac{\partial^2 f}{\partial x^2}(x,z) \\ &+ \sum_{i=1}^{m}\{2\frac{\partial^2 f}{\partial x \partial z_i}(x,z)f_i(x,z) \\ &+ \sum_{j=1}^{m} \frac{\partial^2 f}{\partial z_i \partial z_j}(x,z)f_i(x,z)f_j(x,z) \\ &+ \frac{\partial f}{\partial z_i}(x,z)[\frac{\partial f_i}{\partial x}(x,z) + \sum_{j=1}^{m} \frac{\partial f_i}{\partial z_j}(x,z)]\}, \\ &\ldots\ldots\ldots\ldots\ldots\ldots \end{aligned} \tag{1.3.2}$$

From (1.3.1) and (1.3.2) the different discrete methods are followed. Accepting a uniform net on the interval [a,b] and using z'(x) only, we have

$$\begin{aligned} \eta^o &= z^o, \\ \eta^{k+1} &= \eta^k+hf(x_k,\eta^k); \quad k=0(1)n-1; \end{aligned} \tag{1.3.3}$$

where $\eta^k=\eta(x_k)$, $x_k=x_o+kh$, $h=(b-a)/n$. Of course, this is the Euler method (compare (1.2.3)). If we take also into account the second derivative in (1.3.1), then we obtain a method of the form

$$\begin{aligned} \eta^o &= z^o, \\ \eta^{k+1} &= \eta^k+hf(x_k,\eta^k)+ \frac{h^2}{2}[\frac{\partial f}{\partial x}(x_k,\eta^k) \\ &+ \sum_{i=1}^{m} \frac{\partial f}{\partial z_i}(x_k,\eta^k)f_i(x_k,\eta^k)]; \\ &\qquad k=0(1)n-1. \end{aligned} \tag{1.3.4}$$

Using the Taylor formula (1.3.1) we can obtain the discrete methods up to the order s. For example, Euler's method (1.3.3) is of the first order, what has been shown in Sect. 1.2 in the case m=1 (for an arbitrary m a proof is analogous), and the method (1.3.4) is of the second order.

It is not difficult to note that all discrete methods obtained from (1.3.1) are of one-step methods. It means that in order to obtain the solution at the point x_k (k=1(1)n) it is sufficient to know the solution at the previous point. Unfortunately, their use in practice (instead of Euler's method) is severely restricted to the problems for which the higher derivatives of f can be readily obtained. For instance, for the method (1.3.4), m values of f_i (i=1(1)m) and m^2+m values of partial derivatives of the first order must be calculated, and for the method of the third order we ought to calculate f_i, $\partial f_i/\partial x$, $\partial f_i/\partial z_j$ (j=1(1)m) and m^3+m^2+m values of the second derivatives $\partial^2 f_i/\partial x^2$, $\partial^2 f_i/\partial x \partial z_j$, $\partial^2 f_p/\partial z_i \partial z_j$ (i,j,p=1(1)m).

The methods of order s>1 obtained from Taylor's formula are used first and foremost to calculate the starting points in multistep methods (see Sect.1.3.2) on the condition that the form of f(x,z) is not very complicated. The highly recommended paper related to those methods is [5].

The methods following from the Taylor formula can be written in the form (for a uniform net)

$$\begin{aligned} \eta^o &= z^o, \\ \eta^{k+1} &= \eta^k + h\Phi(h,x_k,\eta^k); \quad k=0(1)n-1; \end{aligned} \tag{1.3.5}$$

where η^k, $\Phi(h,x_k,\eta^k) \in R^m$. For instance, for Euler's method (1.3.3) we have $\Phi(h,x_k,\eta^k)=f(x_k,\eta^k)$.

Around the turn of the century Runge and subsequently Heun and Kutta proposed the methods of the form (1.3.5) which are equivalent to constructing a formula for Φ that agrees as closely as possible with the expression

$$\Delta(h,x,z)=z'(x)+\frac{h^2}{2!}z''(x)+\ldots+\frac{h^{s-1}}{s!}z^{(s)}(x) + R(x,h) \tag{1.3.6}$$

without involving derivatives of f. In this purpose, the function Φ is determined from the formulas

$$\Phi(h,x,z) = \sum_{p=1}^{r} c_p K^p,$$

$$K^p = K^p(h,x,z) = f(x+ha_p, z+h \sum_{q=1}^{r} b_{pq}K^q); \qquad (1.3.7)$$
$$p=1(1)r;$$

where the coefficients a_p, b_{pq} and c_p $(p,q=1(1)r)$ are calculated in such a manner that Φ and Δ ought to be the same up to the terms of possible high order with respect to h. Let us note that if $r>p-1$, then the second equation (1.3.7) yields in general a nonlinear system of equations with respect to K^p. Such methods are called r-stage implicit Runge--Kutta methods. However, if $r=p-1$, then from this equation we obtain a system of linear recurrence equations with respect to the variable mentioned above. These methods are said to be r-stage explicit Runge-Kutta methods.

Since in the implicit Runge-Kutta methods more parameters are available, such formulas are potentially more accurate than the corresponding explicit ones. A disadvantage of implicit methods is that they require a system of nonlinear equations to be solved at each step. Therefore, they are only used in some special problems in which their properties result in the better approximation of the solution than the explicit methods. Further we will only consider the explicit Runge-Kutta methods. The reader can find more details on the implicit methods e.g. in [12].

For the explicit methods the system (1.3.7) has the form

$$\Phi(h,x,z) = \sum_{p=1}^{r} c_p K^p,$$
$$K^1 = f(x,z), \qquad (1.3.8)$$
$$K^p = f(x+ha_p, z+h \sum_{q=1}^{p-1} b_{pq}K^q); \quad p=2(1)r.$$

As examples we work out the formulas of explicit method with $r=1,2$ and 4.

It is easy to check that in the case where $r=1$ we obtain the Euler method. Namely, from (1.3.8) we have

$$\Phi(h,x,z) = c_1 f(x,z)$$

and a comparision with (1.3.6) yields $c_1=1$.

For $r=2$ from (1.3.6) we obtain

$$\Phi(h,x,z) = c_1 f(x,z)+c_2 f[x+ha_2, z+hb_{21}f(x,z)]. \qquad (1.3.9)$$

Expanding the second term, we have

$$f[x+ha_2,z+hb_{21}f(x,z)] = f(x,z) + h[a_2 \frac{\partial f}{\partial x}(x,z)+b_{21} \sum_{i=1}^{m} \frac{\partial f}{\partial z_i}(x,z)f_i(x,z)] + O(h^2).$$

Taking into account (1.3.2), from (1.3.6) and (1.3.9) we get

$$\Delta(h,x,z) - \Phi(h,x,z) = (1-c_1-c_2)f(x,z) + h[(\tfrac{1}{2} -a_2c_2)\frac{\partial f}{\partial x} +(\tfrac{1}{2} -b_{21}c_2) \sum_{i=1}^{m} \frac{\partial f}{\partial z_i}f_i]+O(h^2).$$

Hence, the coefficients ought to fulfil the system of equations of the form

$$\begin{aligned} c_1 + c_2 &=1, \\ a_2c_2 &= 1/2, \\ b_{21}c_2 &= 1/2. \end{aligned} \qquad (1.3.10)$$

So, we have obtained the system of three equations with four unknowns and we get the following set of solutions

$$c_1 = 1-\alpha, \quad c_2 = \alpha, \quad a_2 = b_{21} = 1/2\alpha,$$

wher $\alpha\neq 0$ denotes an arbitrary parameter. For $\alpha=1/2$ we have

$$\Phi(h,x,z) = f(x,z)/2 + f[x+h,z+hf(x,z)]/2.$$

This is Heun's method that yields the formulas

$$\begin{aligned} \eta^o &= z^o, \\ \eta^{k+1} &= \eta^k+h\{f(x_k,\eta^k)+f[x_{k+1},\eta^k+hf(x_k,\eta^k)]\}; \\ & \qquad k=0(1)n-1. \end{aligned}$$

For $\alpha=1$ from (1.3.10) we obtain the modified method of Euler in which

$$\Phi(h,x,z) = f[x+h/2,z+hf(x,z)/2]$$

and

$$\begin{aligned} \eta^o &= z^o, \\ \eta^{k+1} &= \eta^k+hf[x_k+h/2,\eta^k+hf(x_k,\eta^k)/2]; \quad k=0(1)n-1. \end{aligned}$$

Analogous considerations for r=4 bring us to the following equations for coefficients

$$\begin{aligned} c_1+c_2+c_3+c_4 &= 1, \\ c_2a_2+c_3a_3+c_4a_4 &= 1/2, \\ c_2a_2^2+c_3a_3^2+c_4a_4^2 &= 1/3, \\ c_3a_2b_{32}+c_4(a_2b_{42}+a_3b_{43}) &= 1/6, \end{aligned}$$

$$\begin{aligned}
&c_2a_2^3+c_3a_3^3+c_4a_4^3 = 1/4, \\
&c_3a_2^2b_{32}+c_4(a_2^2b_{42}+a_3^2b_{43}) = 1/12, \\
&c_3a_2a_3b_{32}+c_4(a_2b_{42}+a_3b_{43})a_4 = 1/8, \\
&c_4a_2b_{32}b_{43} = 1/24, \\
&a_i = \sum_{j=1}^{i-1} b_{ij}; \quad i=2,3,4.
\end{aligned} \tag{1.3.11}$$

The system (1.3.11) consists of eleven equations with thirteen unknowns. It means that we can calculate a solution with two parameters. However, a set of solution with one parameter is considered most often. In this case we have

$$\begin{aligned}
&c_1 = c_4 = 1/6, \quad c_2 = (2-\alpha)/3, \quad c_3 = \alpha/3, \\
&a_2 = a_3 = b_{21} = 1/2, \quad a_4 = 1, \\
&b_{31} = (\alpha-1)/2\ , \quad b_{32} = 1/2\alpha, \\
&b_{41} = 0, \quad b_{42} = 1-\alpha, \quad b_{43} = \alpha,
\end{aligned} \tag{1.3.12}$$

where $\alpha \neq 0$ is an arbitrary parameter.

For $\alpha=1$ from (1.3.12) we obtain the classical Runge-Kutta method of fourth order in which

$$\Phi(h,x,z) = (K^1+2K^2+2K^3+K^4)/6,$$

where

$$\begin{aligned}
K^1 &= f(x,z), \\
K^2 &= f(x+h/2,z+hK^1/2), \\
K^3 &= f(x+h/2,z+hK^2/2), \\
K^4 &= f(x+h,z+hK^3),
\end{aligned}$$

i.e.

$$\begin{aligned}
\eta^o &= z^o, \\
\eta^{k+1} &= \eta^k+h[K^{1(k)}+2K^{2(k)}+2K^{3(k)}+K^{4(k)}]/6; \quad k=0(1)n-1;
\end{aligned}$$

where

$$\begin{aligned}
K^{1(k)} &= f(x_k,\eta^k), \\
K^{2(k)} &= f[x_k+h/2,\eta^k+hK^{1(k)}/2] = f[x_k+h/2,\eta^k+hf(x_k,\eta^k)/2 \\
K^{3(k)} &= f[x_k+h/2,\eta^k+hK^{2(k)}/2] \\
&= f\{x_k+h/2,\eta^k+hf[x_k+h/2,\eta^k+hf(x_k,\eta^k)/2]/2\}, \\
K^{4(k)} &= f[x_k+h,\eta^k+hK^{3(k)}] \\
&= f(x_k+h,\eta^k+hf\{x_k+h/2,\eta^k+hf[x_k+h/2, \\
&\qquad \eta^k+hf(x_k,\eta^k)/2]/2\}).
\end{aligned}$$

It can be shown that the maximal order for the method of Runge-Kutta for r=1,2,3,4 is equal to r, and for r>4 is not greater than r (e.g. see [32],[47],[86]). A detailed analysis for the Runge-Kutta methods, which is omited here, is given in [15]. These methods are also considered in [12],[13],[20], [24],[25],[26],[27],[61] and [77].

Now, we deal with the practical realization of one-step methods on digital computers. In different problems of celestial mechanics we usually want to obtain the solution of (1.1.1)-(1.1.2) at the equidistant points (time moments). It means that we set some uniform net on the interval [a,b] beforehand. Let h denote a step of such net. Using the one-step method of order p, it can appear that for this step it is impossible to obtain a solution with the accuracy ε given beforehand or that it is possible to get such solution taking a greater step.

Because we want to obtain the solution at equidistant points and with an accuracy given beforehand, if it appears that the step size is too small, then we usually apply a method of lower order (for instance, instead of the classical Runge-Kutta method, the Runge-Kutta method of the second or third order). There is much to be said for the decrease of calculation expenditure.

Another situation arises when the step size h is too great to obtain the solution with the accuracy ε. In this case it is necessary to decrease the step size. It seems to be obvious in this purpose to divide each interval $[x_k, x_k+h]$ $(k=0(1)n-1;\ x_o=a,\ x_n=b)$ on k_ν subintervals and use the method with the step h_k on each such subinterval. However, the problem of choice of the step size still remains, because we must choose h_k in each subinterval. Moreover, it can become evident that the uniform net is not the best in the interval $[x_k, x_k+h]$.

Before we describe the way of integration of the initial value problem with the choice of step size which insures the obtainment of solution with an accuracy given beforehand, let us introduce the notion of an asymptotic expansion of total discretization error (see e.g.[54],[86],[87]). We incidentally return to the notations introduced in Sect. 1.2.

<u>Definition 1.6.</u> Let us consider the discretization method $M=\{E_n, E_n^o, \Delta_n, \Delta_n^o, \phi_n\}_{n\in N'}$, $N'\subset N$, of the order p used in the initial value problem B given by (1.1.6). We say that the total discreti-

zation error $\{\varepsilon_n\}_{n\in N'}$ of M has an asymptotic expansion up to the order J on B, if there exist the elements $e_j \in E$ $(j=p(1)J)$ independent of n such that

$$\varepsilon_n = \Delta_n\left(\sum_{j=p}^{J} \frac{1}{n^j} e_j\right) + O[n^{-(J+1)}] \text{ as } n\to\infty. \tag{1.3.13}$$

The existance of an asymptotic expansion of total discretization error for a lot of discretization methods has been proved in [86]. In particular, one-step methods presented in this section have such expansions. In a general case the following theorem can be proved.

Theorem 1.3. Let the discretization method $M=\{E_n, E_n^o, \Delta_n, \Delta_n^o, \phi_n\}_{n\in N'}$ be applied to the initial value problem $B=\{E, E^o, F\}$ with the exact solution z. Let us assume that M and B fulfil the conditions:

(i) M is stable on B;

(ii) M is consistent with B with order p;

(iii) there exist the mappings Λ_n: $E\to E^o$ such that for each y belonging to a domain of F,

$$\phi_n(F)\Delta_n y = \Delta_n^o(F+\Lambda_n)y,$$

there exist the nonempty subset $D_J \subset E$ and the mappings λ_j: $D_J \to E^o$ $(j=1(1)J)$ independent of n such that for each $y\in D_J$

$$\left|\left| \Delta_n^o\left(\Lambda_n y - \sum_{j=1}^{J} \frac{1}{n^j}\lambda_j y\right)\right|\right|_{E_n^o} = O[n^{-(J+1)}] \quad \text{as } n\to\infty$$

and $z\in D_J$;

(iv) the mapping F has Frèchet's derivatives up to the order [J/p] ([.] denotes the entier) that satisfy the Lipschitz condition in some domain of the form $B_R=\{y\in E: ||y-z||_E < R, R>0\}$;

(v) there exist the derivatives on the left-hand side of the equality

$$\sum_{j=1}^{J} \frac{1}{n^j}\left[\lambda_j y + \sum_{m=1}^{[(J-j)/p]} \frac{1}{m!}\lambda_j^{(m)}(y)\left(\sum_{k=p}^{J}\frac{1}{n^k} e_k\right)^m\right]$$
$$= \Lambda_n\left(y+\sum_{k=p}^{J}\frac{1}{n^k} e_k\right) + O[n^{-(J+1)}] \tag{1.3.14}$$

and this equality holds for the arbitrary $y \in D_J$, $e_k \in D_J$ $(k=p(1)J)$;

(vi) there exists $[F'(z)]^{-1}$.

Then there exist the elements e_j $(j=p(1)J)$ such that M has the unique asymptotic expansion of total discretization error up to the order J on B and this expansion is given by (1.3.13).

The proof of this theorem can be found in [86].

We now describe the way of step size choice ([87]). At first, let us devide the interval [a,b] on the equal subintervals with size h, i.e. $x_k = x_o + kh = a + kh$ $(k=0(1)n)$. In other words, we are interested in the obtainment of the solution of the initial value problem at the points x_k $(k=1(1)n)$ with the accuracy $\bar{\varepsilon}$ given beforehand. Moreover let us assume that the order of method is given and equals p. For simplicity, let us take the first subinterval, i.e. $[a,a+h]=[x_o,x_1]$.

If $\zeta(x,h)$ is a solution of a discrete problem which is consistent with the initial value problem and z is the exact solution of (1.1.1)-(1.1.2), then from (1.3.13) we have

$$\varepsilon(x_1,h) = \zeta(x_1,h)-z(x_1) = h^p e_p(x_1)+O(h^{p+1}) \tag{1.3.15}$$

Let $\zeta(x_1,h/2)$ denote a solution of a discrete problem at the point x_1 obtained with the step h/2. (1.3.15) implies that

$$\begin{aligned} \zeta(x_1,h)-z(x_1) &\doteq h^p e_p(x_1), \\ \zeta(x_1,h/2)-z(x_1) &\doteq (h/2)^p e_p(x_1), \end{aligned} \tag{1.3.16}$$

where $\doteq$ denotes the equality in the first approximation. Subtracting the equalities (1.3.16) we obtain

$$\zeta(x_1,h)-\zeta(x_1,h/2) \doteq (h/2)^p(2^p-1)e_p(x_1),$$

i.e.

$$(h/2)^p e_p(x_1) \doteq \frac{\zeta(x_1,h)-\zeta(x_1,h/2)}{2^p-1} .$$

Hence, after the substitution in the second relationship of (1.3.16), we have

$$\zeta(x_1,h/2)-z(x_1) \doteq \frac{\zeta(x_1,h)-\zeta(x_1,h/2)}{2^p-1} . \tag{1.3.17}$$

Now, let $\bar{\varepsilon}$ denote a desired accuracy of a solution. Let us try to determine the size of $\tilde{h}=\tilde{h}(\varepsilon)$ so as to have the discretization error at the point $x_o+\hbar$ equal $\bar{\varepsilon}$ approximately, i.e.

$$||\ \varepsilon(x_o+\hbar,\hbar)\ ||\ \simeq \bar{\varepsilon}, \qquad (1.3.18)$$

where $||\varepsilon(.,.)|| = \max_{i=1(1)m} \varepsilon_i(.,.)$ and m is a dimension of the vector ε. For a method of order p we have

$$\varepsilon(x_o+\hbar,\hbar) \doteq \hbar^p e_p(x_o+\hbar). \qquad (1.3.19)$$

But

$$e_p(x_o+\hbar) = e_p(x_o)+\hbar e_p'(x_o)+O(\hbar^2).$$

Moreover, $e_p(x_o)=0$, and so

$$e_p(x_o+\hbar) \doteq \hbar e_p'(x_o). \qquad (1.3.20)$$

Consequently, the condition (1.3.18) is fulfiled if

$$\bar{\varepsilon} \doteq \hbar^p||e_p(x_o+\hbar)|| \doteq \hbar^{p+1}||e_p'(x_o)||. \qquad (1.3.21)$$

Let us calculate $e_p'(x_o)$. Let H denote some given integration step (do not mistake H and h). From (1.3.17) we have

$$\varepsilon(x_o+H,H/2) \doteq \frac{\zeta(x_o+H,H)-\zeta(x_o+H,H/2)}{2^p-1}. \qquad (1.3.22)$$

In view of (1.3.19) and (1.3.20) we obtain

$$\varepsilon(x_o+H,H/2) \doteq (H/2)^p e_p(x_o+H) \doteq \frac{H^{p+1}}{2^p} e_p'(x_o).$$

Hence, taking into account (1.3.22), we get

$$e_p'(x_o) \doteq \frac{2^p}{2^p-1}\frac{1}{H^{p+1}}[\zeta(x_o+H,H)-\zeta(x_o+H,H/2)].$$

Finally, (1.3.21) yields the following integration step formula

$$\hbar \doteq H/\sqrt[p+1]{\frac{2^p}{2^p-1}\ \frac{\zeta(x_o+H,H)-\zeta(x_o+H,H/2)}{\bar{\varepsilon}}}. \qquad (1.3.23)$$

The formula (1.3.23) is applied as follows: We choose H (for the first integration step it is advisable to set H=h), substitute $t=x_o$, calculate $\zeta(x_o+H,H)$, $\zeta(x_o+H,H/2)$ and $\hbar$ from (1.3.23). If $\hbar \ll H/2$, then (1.3.22) implies that $\varepsilon(x_o+H,H/2) \gg \bar{\varepsilon}$. It means that the step size must be decreased. We replace H by $2\hbar$ and for this new value of H we calculate $\zeta(x_o+H,H)$, $\zeta(x_o+H,H/2)$ and $\hbar$ once more. We proceed in such a way as long as $\tilde{h}<<H/2$. If $\tilde{h}\simeq H/2$, then we accept $\zeta(x_o+H,H/2)$ as an approximation of the exact solution $z(x_o+H)$, increase t to t_o+H and then, replacing the previous initial values by t and $z(t)$ and H by $2\tilde{h}$, we begin the next integration step.

Since we are anxious to obtain a solution at the point x_o+h, we ought to make sure that $t\leq x_o+h$ each time when t is increased. If it appears that $t>x_o+h$, i.e. we cross the point x_o+h, then the actual value of H ought to be memorized (in order to begin the integration procedure in the interval $[x_o+h,x_o+2h]$), and the next integration step must be repeated with the step $x_o+h-(t-H)$ which insures that we get the solution at the point x_o+h.

If the condition $\tilde{h}>>H/2$ is fulfiled instead of $\tilde{h}<<H/2$, then from (1.3.22) it follows that $\varepsilon(x_o+H,H/2)<<\bar{\varepsilon}$. In this case we can increase the step size replacing H by $2H$ or we can integrate further with the step H, although the second variant increases the calculation expenditure unnecesarily. One ought to check to be sure that the point x_o+h is not crossed.

The procedure described above applies so long as we obtain the solution at each point $x_k=x_o+kh$ $(k=1(1)n)$. Moreover, it ought to be noted that the solution accuracy $\bar{\varepsilon}$ given beforehand is related to the accuracy of a digital machine and must fulfil the condition

$$\bar{\varepsilon} \geq 5\times 10^{-\tau}||\ z\ ||_E,$$

where τ is the number of digits of the machine number representation and z denotes, as usual, the exact solution of (1.1.6). In practice the conditions $\tilde{h}<<H/2$ and $\tilde{h}>>H/2$ are replaced by $\tilde{h}<H/3$ and $\tilde{h}>H$ respectively. The algorithm of an automatic step size correction for solving the initial value problem by an arbitrary one-step method of order p is illustrated in Fig.2.

To end with this section we would like to notice the effect of rounding (round-off) errors on the obtained solution. Denoting the exact solution of the discretization method by ζ_n and the solution obtained in τ-digital floating-point arithmetic by $\bar{\zeta}_n$, we can get the following estimation for the rounding error (see e.g.[54],[87])

$$|| \zeta_n-\bar{\zeta}_n ||_{E_n} \leq n\bar{\varepsilon}\,\frac{e^L-1}{L}, \qquad (1.3.24)$$

where n=1/h. In (1.3.24) it is assumed that all rounding errors are equal to $\bar{\varepsilon}$ and that the function Φ, which defines one-step method (see (1.3.5)), fulfils the Lipschitz condition of the form

$$|| \Phi(h,x_k,\eta^k)-\Phi(h,x_k,\bar{\eta}^k) || \leq L|| \eta^k-\bar{\eta}^k ||.$$

The estimation (1.3.24), although not over-precise enough to be of practical importance, yields us the significant particulars. Namely, in the τ-digital floating-point arithmetic we can obtain the solution which has a large error while the step h is small. This means that the step size can be decreased only to some sensible proportions. Assuming that the initial condition is given exactly, from Def.1.4 and (1.3.24) we get for a method of order p the following estimation

$$|| \bar{\zeta}_n-\Delta_n z || \leq || \zeta_n-\Delta_n z || + || \zeta_n-\bar{\zeta}_n ||$$
$$\leq An^{-p}+Bn = Ah^p+B/h,$$

where A and B denote constants. From the above inequality it is self-evident that the total error increases as $h\to 0$ as well as $h\to\infty$ (see also Fig.3).

1.3.2. LINEAR MULTISTEP METHODS

Before we present the general definition of the multistep method, let us consider a construction example for such a method. At first, let us note that the solution z=z(x) of the initial value problem fulfils for each $x, x-qh \in [a,b]$ an integral equation of the form

$$z(x) = z(x-qh) + \int_{x-qh}^{h} f[t,z(t)]dt. \qquad (1.3.25)$$

To carry out integration in (1.3.25), we can approximate f[t,z(t)] by Lagrange's interpolation polynomial L_μ of the degree not greater than μ which interpolates our function at the points $x_l=a+lh$ $(l=\nu-p(-1)\nu-p-\mu,\ h=(b-a)/n)$, i.e.

$$L_\mu(t) = \sum_{i=1}^{\mu} f[x_{\nu-p-i},z(x_{\nu-p-i})]s_i(t), \qquad (1.3.26)$$

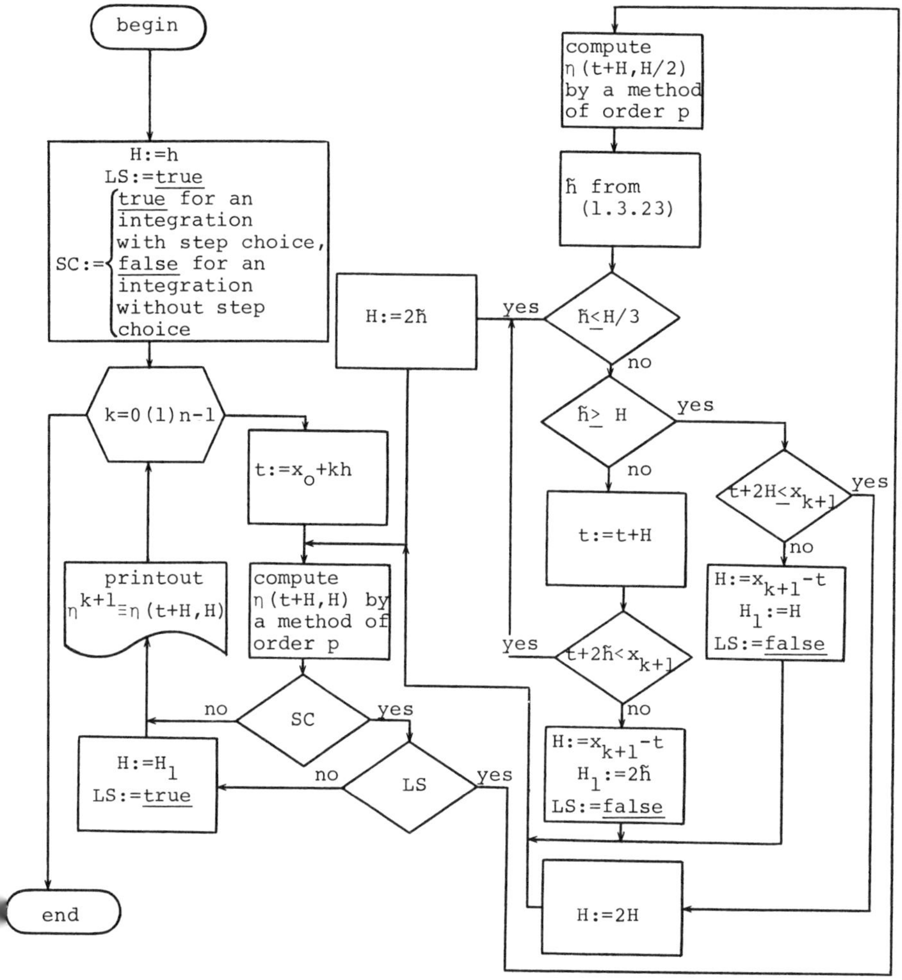

Fig.2. Solution of the initial value problem with an automatic step size correction by user's desire. (It is assumed that the step size can not be greater than h.)

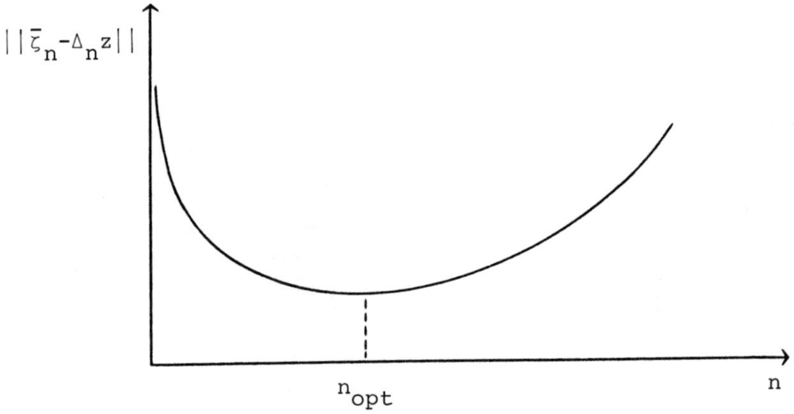

Fig.3. Optimal step size (n_{opt} denotes the value that gives $\min|| \bar{\zeta}_n - \Delta_n z ||_{E_n}$)

where

$$s_i(t) = \prod_{\substack{r=0 \\ r\neq i}}^{\mu} \frac{t - x_{\nu-p-r}}{x_{\nu-p-i} - x_{\nu-p-r}} . \tag{1.3.27}$$

Since $L_\mu(t) \simeq f[t, z(t)]$, then from (1.3.25)-(1.3.27) we get

$$\eta^\nu = \eta^{\nu-q} + \sum_{i=0}^{\mu} f(x_{\nu-p-i}, \eta^{\nu-p-i}) \int_{x_{\nu-q}}^{x_\nu} s_i(t)\,dt, \tag{1.3.28}$$

where $\eta^\nu = \eta(x_\nu)$. To compute the integral which occurs in (1.3.28), let us substitute $t = x_\nu + uh$. We have

$$\eta^\nu = \eta^{\nu-q} + h \sum_{i=0}^{\mu} \gamma_{\mu i} f(x_{\nu-p-i}, \eta^{\nu-p-i}), \tag{1.3.29}$$

where

$$\gamma_{\mu i} = \int_{-q}^{0} \prod_{\substack{r=0 \\ r\neq i}}^{\mu} \frac{u+p+r}{r-i}\, du. \tag{1.3.30}$$

Now, let $\mu = k-p$. Then (1.3.29) yields

$$\eta^\nu = \eta^{\nu-q} + h \sum_{i=0}^{k-p} \gamma_{k-p,i} f(x_{\nu-p-i}, \eta^{\nu-p-i})$$

$$= \eta^{\nu-q} + h \sum_{i=0}^{k-p} \gamma_{k-p,k-p-i} f(x_{\nu-k+i}, \eta^{\nu-k+i}). \quad (1.3.31)$$

Denoting the coefficient $\gamma_{k-p,k-p-i}$ by β_i, from (1.3.31) we obtain conclusively

$$\eta^{\nu} = \eta^{\nu-q} + h \sum_{i=0}^{k-p} \beta_i f(x_{\nu-k+i}, \eta^{\nu-k+i}). \quad (1.3.32)$$

The above formula defines some class of linear k-step methods. Let us note that choosing the parameter p, the relation (1.3.32) presents a linear or nonlinear equation with respect to unknown quantity η^{ν} (explicit and implicit methods) and that in order to compute η^{ν} we must know the values of a solution at most k previous points. These values can be obtained using a so called started procedure which most often is one of the Runge-Kutta methods.

Let us consider a few special examples of the formulas (1.3.29)-(1.3.30). Let the values of the solution at the points x_{ν} ($\nu=0(1)k-1$) be known. For q=1 and p=1 we obtain an explicit method of Adams-Bashforth of the form ($h=(b-a)/n$)

$$\eta^{\nu} = \eta^{\nu-1} + h \sum_{i=0}^{k-1} \gamma_{k-1,i} f(x_{\nu-i-1}, \eta^{\nu-i-1}); \quad \nu=k(1)n. \quad (1.3.33)$$

Accepting q=1 and p=0 we have an implicit method of Adams--Moulton in which

$$\eta^{\nu} = \eta^{\nu-1} + h \sum_{i=0}^{k} \gamma_{k,i} f(x_{\nu-i}, \eta^{\nu-i}); \ \nu=k(1)n. \quad (1.3.34)$$

For q=2 and p=1 from (1.3.29) we get an explicit method of Nyström of the form

$$\eta^{\nu} = \eta^{\nu-2} + h \sum_{i=0}^{k-1} \gamma_{k-1,i} f(x_{\nu-i-1}, \eta^{\nu-i-1}); \quad \nu=k(1)n. \quad (1.3.35)$$

Next, accepting q=2 and p=0, we obtain the following method of Milne

$$\eta^{\nu} = \eta^{\nu-2} + h \sum_{i=0}^{k} \gamma_{k,i} f(x_{\nu-i}, \eta^{\nu-i}); \ \nu=k(1)n. \quad (1.3.36)$$

In practice the parameter $\mu=k-p$ is not greater than six. In Tables I-IV the values of coefficients $\gamma_{\mu,i}$ and the orders of methods (1.3.33)-(1.3.36) are presented.

Table I. Coefficients $\gamma_{\mu,i}=\rho_{\mu,i}/\sigma_\mu$ for the method of Adams--Bashforth (J denotes the order)

	$\rho_{\mu,i}$								
$\mu \backslash i$	0	1	2	3	4	5	6	σ_μ	J
0	1							1	1
1	3	-1						2	2
2	23	-16	5					12	3
3	55	-59	37	-9				24	4
4	1901	-2774	2616	-1274	251			720	5
5	4227	-7673	9482	-6798	2627	-425		1440	6
6	198721	-447288	705549	-688256	407139	-134472	19087	60480	7

Table II. Coefficients $\gamma_{\mu,i}=\rho_{\mu,i}/\sigma_\mu$ for the method of Adams--Moulton

	$\rho_{\mu,i}$								
$\mu \backslash i$	0	1	2	3	4	5	6	σ_μ	J
0	1							1	1
1	1	1						2	2
2	5	8	-1					12	3
3	9	19	-5	1				24	4
4	251	646	-264	106	-19			720	5
5	475	1427	-798	482	-173	27		1440	6
6	19087	65112	-46461	37504	-20211	6312	-863	60480	7

Table III. Coefficients $\gamma_{\mu,i}=\rho_{\mu,i}/\sigma_\mu$ for the method of Nyström

	$\rho_{\mu,i}$								
$\mu \backslash i$	0	1	2	3	4	5	6	σ_μ	J
0	2							1	2
1	2	0						1	2
2	7	-2	1					3	3
3	8	-5	4	-1				3	4
4	73	-130	174	-106	25			18	5
5	297	-406	574	-426	169	-28		90	6
6	13613	-23886	41193	-40672	24183	-8010	1139	3780	7

Table IV. Coefficients $\gamma_{\mu,i}=\rho_{\mu,i}/\sigma_\mu$ for the method of Milne

	$\rho_{\mu,i}$							σ_μ	J
$\mu \backslash i$	0	1	2	3	4	5	6		
0	1							1	1
1	0	2						1	2
2	1	4	1					3	4
3	1	4	1	0				3	4
4	29	124	24	4	-1			90	5
5	28	129	14	14	-6	1		90	6
6	1139	5640	33	1328	-807	264	-37	3780	7

For $\mu=k=1$ from (1.3.34) we obtain

$$\eta^\nu = \eta^{\nu-1} + \frac{h}{2}[f(x_\nu,\eta^\nu)+f(x_{\nu-1},\eta^{\nu-1})]; \quad \nu=1(1)n. \qquad (1.3.37)$$

This formula is called the trapezium rule. Next, for $\mu=0$, i.e. $k=1$, from (1.3.35) we get **the** so called central difference formula of the form

$$\eta^\nu = \eta^{\nu-2} + 2hf(x_{\nu-1},\eta^{\nu-1}); \quad \nu=2(1)n; \qquad (1.3.38)$$

which is also known as the midpoint method or rule.

Let us note that the accuracy order of (1.3.38) is the same as for Nyström's method with $k=2$ (see Table III). An analogous situation occurs for Milne's method. The accuracy order of (1.3.36) is the same for $\mu=2$ and $\mu=3$ (see Table IV).

The formulas (1.3.35) and (1.3.36) "jump over" one point and this implies that the order of accuracy is increased. However, on the other hand, this causes a problem of weak stability to come into being. Such weak stability can generate, under some circumstances, a "parasitical" solution which perturbs the solution obtained by Nyström's or Milne's method. More extensive analysis of the problem of weak stability can be found e.g. in [68] and [69].

The formula (1.3.32) determines some class of linear multistep methods whose general definition, using the notation from Sect.1.2, is as follows:

Definition 1.7. The discretization method $M=\{E_n, E_n^o, \Delta_n, \Delta_n^o, \phi_n\}$ applied to the initial value problem

$B=\{E,E^o,F\}$ of the form (1.1.6) is called a linear k-step method if the net G_k is uniform and

$$[\phi_n(F)\eta](\tfrac{\nu}{n}) = \begin{cases} \sum_{i=1}^{k} [\frac{\alpha_i\eta^{\nu-k+i}}{1/n} - \beta_i f(x_{\nu-k+i}, \eta^{\nu-k+i})], & \nu=k(1)n, \\ \eta^\nu - s^{n\nu}(z^o), & \nu=0(1)k-1, \end{cases} \qquad (1.3.39)$$

where $\eta^\nu=\eta(x_\nu)$, $\sum_{i=0}^{k} \alpha_i = 0$, $\alpha_k \neq 0$, $\alpha_o^2+\beta_o^2>0$, $n=1/h$. The terms $[\phi_n(F)\eta](\nu/n)$ are called a linear k-step procedure if $\nu=k(1)n$, and one said to be a starting procedure if $\nu=0(1)k-1$.

As we noted previously, as a starting procedure we use most often the method of Runge-Kutta and then

$$s^{n\nu}(z^o) = \begin{cases} z^o, & \nu=0, \\ \eta^{\nu-1} - \frac{1}{n} \sum_{p=1}^{r} c_p K^p, & \nu=1(1)k-1, \end{cases} \qquad (1.3.40)$$

where the K^p are given by (1.3.7). In practice, to calculate (1.3.40) the value $\bar{n}>>n$ is used instead of n or it is applied an automatic step size correction that gives the values of (1.3.40) with the accuracy given beforehand (see the previous section).

The reader can verify that the formula (1.3.32) is a special case of (1.3.39).

For multistep methods the norm in the space E_n^o we define as follows (compare with (1.2.1))

$$||\ \delta\ ||_{E_n^o} = \max_{\nu=0(1)k-1} (\sum_{i=1}^{m} |\delta_i(\tfrac{\nu}{n})|) + \frac{1}{n} \sum_{\nu=k}^{n} \sum_{i=1}^{m} |\delta_i(\tfrac{\nu}{n})|.$$

If we take a norm as above, then from Defs.1.3 and 1.5 we can obtain the definitions of consistency and stability for linear multistep methods most often met in references (see e.g. [32],[44],[47],[60],[68],[69]).

<u>Definition 1.8.</u> Let us create the expressions

$$C_o = \alpha_o+\alpha_1+\ldots+\alpha_k,$$
$$C_1 = \alpha_1+2\alpha_2+\ldots+k\alpha_k-(\beta_o+\beta_1+\ldots+\beta_k),$$

$$C_2 = \frac{1}{2!}(\alpha_1+2^2\alpha_2+\ldots+k^2\alpha_k) - \frac{1}{1!}(\beta_1+2\beta_2+\ldots+k\beta_k),$$
$$\ldots\ldots\ldots\ldots\ldots\ldots \qquad (1.3.41)$$
$$C_s = \frac{1}{s!}(\alpha_1+2^s\alpha_2+\ldots+k^s\alpha_k) - \frac{1}{(s-1)!}(\beta_1+2^{s-1}\beta_2+\ldots+k^{s-1}\beta_k).$$

We say that a linear k-step procedure is of the order p if $C_o=C_1=\ldots=C_p=0$ and $C_{p+1}\neq 0$. If this procedure is of the order greater or equals 1, then it is consistent with the initial value problem B.

Definition 1.9. Let us take into account the polynomial

$$p(z) = \alpha_o+\alpha_1 z+\ldots+\alpha_k z^k,$$

which is related to (1.3.39) by the coefficients α_i $(i=0(1)k)$. We say that a linear k-step procedure is stable if the polynomial p(z) has only the roots whose absolute value is not greater than 1 and the roots whose absolute value is equal to 1 are single at the same time.

It appears that:

Theorem 1.4. Consistency and stability are together necessary and sufficient conditions for convergence of any linear k-step procedure.

The proof of this theorem can be found e.g. in [47].

Let us note that according to the value of β_k, from (1.3.39) we obtain an explicit or implicit relationship with respect to η^ν $(\nu=k(1)n)$. Therefore, we can distinguish two classes of linear multistep methods.

Definition 1.10. A linear k-step method is said to be explicit if $\beta_k=0$. Otherwise this method is called implicit.

Taking into account the condition $\alpha_k\neq 0$ (see Def.1.7), it is self-evident that each explicit linear k-step procedure can be solved with respect to η^ν. For an implicit procedure a unique solution exists only for sufficiently small step size 1/n. We have for

Theorem 1.5. There exists $n_o>0$ such that an implicit k-step linear procedure has a unique solution $\eta^{\nu}\epsilon B_R(z) =\{y\epsilon E: ||y-z||_E<R, R>0\}$ for all $n>n_o$ and $\eta^{\nu-k+i}\epsilon B_o\equiv B_{R/2a}(z)$ $(i=0(1)k-1)$, where

$$a = \frac{1}{|\alpha_k|}\sum_{i=0}^{k-1}|\alpha_i|.$$

The proof of this theorem occurs in [86]. Further interesting details related to multistep methods and their properties can be found, among others, in [21],[33],[47],[52] and [58].

The following question comes into being: what are implicit methods considered for if they present an obligation to solve a nonlinear system of equations in each step? It appears (see e.g.[32]) that in general the implicit methods have better numerical properties than explicit ones. In the first place we ought to make mention of the following properties:

(i) better estimation of error (for methods of the same order);

(ii) higher accuracy of obtained solution.

Concatenation of both methods into one so called predictor--corrector method gives the desired numerical results.

Since implicit methods require one to solve a nonlinear system of equations in each step, then using an iteration process, we must determine an initial approximation. This approximation ought to be as good as possible in order to obtain the smallest number of iterations that can be gotten from an explicit method of similar order. This way we obtain a predictor-corrector method in which an explicit method is a predictor and an implicit one is a corrector.

As we have mentioned above, to solve a nonlinear system of equations we use an iteration process. From the point of the number of calculations (in other words: in connection with the desired speed of computations), it is advisable to apply some variant of Newton's method described below.

Let us consider a nonlinear system of equations of the form

$$F(y) = 0, \tag{1.3.42}$$

where $y\epsilon R^m$. Let $y^{(s)}$ denote s-th approximation of the solution y of (1.3.42). (s+1)-st approximation is calculated from the relations as follows

$$y_1^{(s+1)} = y_1^{(s)} - \omega \frac{F_1[y_1^{(s)},y_2^{(s)},\ldots,y_m^{(s)}]}{\frac{\partial F_1}{\partial y_1}[y_1^{(s)},y_2^{(s)},\ldots,y_m^{(s)}]},$$

$$y_2^{(s+1)} = y_2^{(s)} - \omega \frac{F_2[y_1^{(s+1)},y_2^{(s)},\dots,y_m^{(s)}]}{\frac{\partial F_2}{\partial y_2}[y_1^{(s+1)},y_2^{(s)},\dots,y_m^{(s)}]},$$

$$\dots\dots\dots\dots\dots\dots \tag{1.3.43}$$

$$y_m^{(s+1)} = y_m^{(s)} - \omega \frac{F_m[y_1^{(s+1)},y_2^{(s+1)},\dots,y_{m-1}^{(s+1)},y_m^{(s)}]}{\frac{\partial F_m}{\partial y_m}[y_1^{(s+1)},y_2^{(s+1)},\dots,y_{m-1}^{(s+1)},y_m^{(s)}]},$$

where the parameter ω fulfils the condition $0<\omega<2$. The value $\omega=1$ is accepted most often. We finish the iteration process (1.3.43) if

$$| y_i^{(s+1)} - y_i^{(s)} | < \varepsilon$$

for each $i=1(1)m$, where ε denotes an accuracy given beforehand.

Let us note that for $\omega=1$ and $m=1$ the method (1.3.43) is the method of Newton for a scalar nonlinear equation. Contrary to Newton's method for a system of nonlinear equations, the process (1.3.43) does not contain the Jacobian $[\partial F_i/\partial y_j]$ $(i,j=1(1)m)$. Moreover, $(s+1)$-st approximations are given explicitly.

For an implicit linear multistep method from (1.3.39) we find easily that

$$F_i(y) = y_i + \sum_{l=0}^{k-1} [\frac{\alpha_l}{\alpha_k} \eta_i^{\nu-k+l} - \frac{1}{n}\frac{\beta_l}{\alpha_k} f_i(x_{\nu-k+l},\eta^{\nu-k+l})]$$
$$- \frac{1}{n}\frac{\beta_k}{\alpha_k} f_i(x_\nu,y)$$

and

$$\frac{\partial F_i}{\partial y_i}(y) = 1 - \frac{1}{n}\frac{\beta_k}{\alpha_k}\frac{\partial f_i}{\partial y_i}(x_\nu,y),$$

where, for simplicity, the solution $\eta^\nu=\eta(\nu/n)$ $(\nu=k(1)n)$ is denoted by y.

Now, let us assume that the initial value problem of the form (1.1.6) we solve by a predictor-corrector method with a strictly determined starting procedure. Let an implicit multistep method of the order p be a corrector. The problem of predictor choice, i.e. the choice of explicit linear multistep method and its order, comes into being. As a predictor we choose most often the explicit method that correspond to a corrector. For instance, for the method of

Adams-Moulton we choose the Adams-Bashforth method and for Milne's method, the method of Nyström. The order of predictor is chosen in such a way to insure the smallest number of iteration for a corrector. Determining this number as two, three or four, we ought to choose a method of the order p or p-1 as a predictor. For futher details we refer e.g. to [32].

For a method of the predictor-corrector type we can consider a problem of step size choice (to obtain a desired accuracy of the solution). However, in predictor-corrector methods with changeable step size we lose the main advantage of linear multistep methods, namely the small number of calculations. Each change of step size creates the necessity to compute additional points of the solution.

The problem of step size change in multistep methods has been considered in a lot of papers. In the instance of Adams's method it is particularly noteworthy to mention the method of Gear ([32]) based on Nordsieck's formulas ([70]) and the effective algorithms of Krogh ([57]) dealing with the cases of increased by twofold step size or the reduction of one by a half.

Among the papers relative to predictor-corrector methods it also ought to be mentioned [19],[34],[43],[46],[55],[56] and [85].

1.3.3. EXTRAPOLATION METHODS

Asymptotic expansions of total discretization error are also applied, except for an automatic step size correction considered in Sect.1.3.1, to the Richardson extrapolation method. An idea of this extrapolation and basic properties is discussed below using the notations from Sect.1.2.

In the Richardson extrapolation we consider the solution ζ_n of the convergent discrete problem as a function of n (we use the step 1/n). If we compute a few values ζ_{n_s} of such function and interpolate these values by the adequate function $\chi=\chi(n)$, then taking $\chi(\infty)$, we obtain an approximation of the exact solution of the initial value problem (1.1.6).

Let us introduce a few indispensable notions ([86]).

<u>Definition 1.11.</u> For the given set I and the space X let $C_r \subset (I \to X)$, $r \in N$, denote such set of functions

determined on I and with values in X that for arbitrary different $n_s \in I$ and $x_s \in X$ $(s=0(1)r)$ given beforehand there exists one and only one function $\chi \in C_r$ which fulfils the condition

$$\chi(n_s)=x_s. \qquad (1.3.44)$$

The sequence $\{C_r\}_{r \in N}$ of such sets is called a class of interpolating functions from I into X.

For example, polynomials establish a class of interpolating functions from R into R. In this case the sets C_r contain the polynomials of the degree not greater than r.

Definition 1.12. A class of interpolating functions is said to be linear if each set C_r is a linear space of the dimension r+1. Otherwise we say that the class is nonlinear.

For the linear class of interpolating functions fulfiling (1.3.44), we have

$$\chi(n) = \sum_{s=0}^{r} \gamma_s(n)x_s. \qquad (1.3.45)$$

Now, let us consider the discretization method $M=\{E_n, E_n^o, \Delta_n, \Delta_n^o, \phi_n\}_{n \in N'}$, $N' \subset N$, applied to the initial value problem $B=\{E, E^o, F\}$. Let this method have an asymptotic expansion of the total discretization error up to the order J (see Def. 1.6). Let us assume that an adequate class of interpolating functions contains the functions for which there exist finite limits as $n \to \infty$. Let ζ_{n_s}, $n_s \in \tilde{N} \subset N'$ $(s=0(1)r)$, denote the solutions of the discrete initial value problem M(B). According to Def.1.11, the elements ζ_{n_s} determine one and only one function $\chi \in C_r$ such that $\chi(n_s)=\zeta_{n_s}$. The Richardson extrapolation relies on the computation of

$$\chi(n) = \lim_{n \to \infty} \chi(n)$$

and the admission of this value as an approximation of the exact solution of the initial value problem B.

In order to accept such a proceeding, the considered class of interpolating functions ought to have the same form of asymptotic expansion with respect to 1/n as the solution ζ_n, i.e. for each function

$$\chi(n) = \text{const} + O(n^{-p}),$$

where p denotes the order of discretization method. If the expansion (1.3.13) contains only the even powers of 1/n, then the functions χ ought also to be even with respect to n.

We must note in this place that some linear multistep methods do not fulfil the condition (1.3.13). However, it appears that for such methods there exists a so called $\hat{\Delta}$-reductable asymptotic expansion, where $\hat{\Delta}$: $E \to \hat{E}$, $||\hat{\Delta}||=1$ and where $\hat{E}$ denotes some given space of Banach. The existance of such expansion is sufficient to apply the Richardson extrapolation. Let us note that the existance of asymptotic expansion of the total discretization error implies the existance of $\hat{\Delta}$-reductable expansion, but the inverse implication is false. For futher details we refer to [86].

In the next passage of this section we will assume that the solution ζ_n of a discrete problem of the order p has an asymptotic expansion of total discretization error up to the order J, i.e. that for $n_s \epsilon \tilde{N} \subset N'$ (s=0,1,...) we have

$$\zeta_{n_s} = \Delta_{n_s} z + \Delta_{n_s} \left(\sum_{j=p}^{J} \frac{1}{n_s^j} e_j \right) + O[n_s^{-(J+1)}] \quad (1.3.46)$$

$$\text{as } n \to \infty,$$

where z is the exact solution of the initial value problem (1.1.6).

Let us denote $T_{s,0} := \zeta_{n_s}$ and let $n_{s+1} > n_s$ (s=0,1,...). Let us take some class $\{C_r\}_{r \epsilon N}$ of interpolating functions on the interval [0,1] and let $\chi_r^l \epsilon C_r$ be such a function that

$$\chi_r^l\left(\frac{1}{n_s}\right) = T_{s,0}; \quad s=l(1)l+r.$$

Let us substitute $T_{lr} := \chi_r^l(0)$. In accordance with Def.1.12, for a linear extrapolation we have

$$T_{lr} = \sum_{s=l}^{l+r} \gamma_{rsl} T_{s,0}, \quad (1.3.47)$$

where the coefficients γ_{rsl} depend on the class $\{C_r\}$ and sequence $\{n_s\}$. We have

<u>Theorem 1.6.</u> Let z be the exact solution of the initial value problem (1.1.6) and let us assume that $\lim_{s \to \infty} T_{s0} = z$.

If

$$\sum_{s=1}^{l+r} \gamma_{rsl} = 1 \quad \text{for each } l \text{ and } r,$$

then $\lim_{l\to\infty} T_{lr}=z$ for each fixed $r>0$.

If, together with the above assumption, the estimation

$$\sum_{s=1}^{l+r} |\gamma_{rsl}| \leq C \tag{1.3.48}$$

holds uniformly with respect to all l and r, and $\lim_{r\to\infty} \gamma_{rsl}=0$ for arbitrary fixed l and $s\geq 1$, then $\lim_{r\to\infty} T_{lr}=z$ for each fixed $l\geq 0$.

The reader can find the proof of this theorem in [86].

In the linear extrapolation a class of polynomials is chosen most often as the class $\{C_r\}_{r\in N}$ and the sequence $\{n_s\}$ is taken in such a way that $n_s=\beta_s n$, where $1/n$ denotes a given basic step size of the solution and $\{\beta_s\}$ is an exactly increasing sequence of natural numbers. For instance, β_s can be defined as follows

$$\beta_s = 2^s; \quad s=0,1,\ldots; \tag{1.3.49}$$

$$\beta_s = s+1; \quad s=0,1,\ldots; \tag{1.3.50}$$

$$\beta_s = \begin{cases} 1, \text{ if } s=0, \\ 2^{(s+1)/2}, \text{ for } s \text{ odd}, \\ 3\times 2^{(s-2)/2}, \text{ for } s \text{ even}. \end{cases} \tag{1.3.51}$$

The sequence $\{\beta_s\}$, where β_s are given by (1.3.49), is called the Romberg sequence and leads to accurate results but is often expensive on a computer as halving the step size invariably doubles the work. The sequence $\{\beta_s\}$, where β_s are given by (1.3.50), is cheaper to compute but leads to an unstable form of the extrapolation algorithm. The sequence in which elements are determined by (1.3.51) is called the Bulirsch sequence and has the advantages of the other two since it leads to a stable algorithm (see [10]) but does not double the cost of the calculations.

If for a discrete method of the order p the quantities $T_{s,0}$

have asymptotic expansions up to the order p+r, i.e. J=p+r (see (1.3.46)) and the polynomials belonging to the class $\{C_r\}$ fulfil the condition

$$\chi'(0) = \ldots = \chi^{(p-1)}(0) = 0, \qquad (1.3.52)$$

then it can be proved (see e.g. [86]).

Theorem 1.7. If $n_s = \beta_s n$ $(s=0(1)r)$, where β_s are fixed and the coefficients γ_{rs0} fulfil the inequality

$$\sum_{s=0}^{r} |\gamma_{rs0}| \leq C$$

uniformly with respect to $n \in N$, then

$$T_{0r} = \Delta_n z + O[n^{-(p+r)}] \quad \text{as } n \to \infty.$$

If the asymptotic expansion (1.3.46) contains only the even powers of 1/n and p=2, then

$$T_{0r} = \Delta_n z + O[n^{-(2r+2)}] \quad \text{as } n \to \infty.$$

From the above theorem it follows that discretization methods having adequate expansions with respect to the even powers of 1/n are of great importance. The method of mid-point (1.3.38) is one of them.

Th.1.7 can also be expressed as

Theorem 1.8. If for a discrete method of the order p the quantities T_{s0} have asymptotic expansions up to the order p+r, $n_s = \beta_s n$ $(s=0(1)r)$, where $\{\beta_s\}$ denotes an exactly increasing sequence of natural numbers, 1/n is a basic step size of the solution and the coefficients γ_{rs} fulfil the relations

$$\sum_{s=0}^{r} \gamma_{rs} = 1,$$
$$\sum_{s=0}^{r} \frac{\gamma_{rs}}{(\beta_s n)^j} = 0, \quad \text{for } j=p(1)p+r-1, \qquad (1.3.53)$$

then

$$T_{0r} = \Delta_n z + O[n^{-(p+r)}] \quad \text{as } n \to \infty. \qquad (1.3.54)$$

Proof. At first, let us note that the system (1.3.53) always has a unique solution. Moreover, from (1.3.46), (1.3.47) and

(1.3.53) we have

$$T_{0r}(\frac{\nu}{n}) = \sum_{s=0}^{r} \gamma_{rs} T_{s0}(\frac{\nu}{n})$$
$$= \sum_{s=0}^{r} \gamma_{rs} [z(\frac{\nu}{n}) + \sum_{j=p}^{p+r} \frac{1}{n_s^j} e_j(\frac{\nu}{n}) + O[n_s^{-(p+r+1)}]$$
$$= z(\frac{\nu}{n}) + \sum_{j=p}^{p+r-1} e_j(\frac{\nu}{n}) \sum_{s=0}^{r} \frac{\gamma_{rs}}{n_s^j} + \sum_{s=0}^{r} \frac{\gamma_{rs}}{n_s^{p+r}} e_{p+r}(\frac{\nu}{n})$$
$$+ O[n_s^{-(p+r+1)}]$$
$$= z(\frac{\nu}{n}) + \sum_{s=0}^{r} \frac{\gamma_{rs}}{n_s^{p+r}} e_{p+r}(\frac{\nu}{n}) + O[n_s^{-(p+r+1)}].$$

Since $n_s \geq n$, the function e_{p+r} is bounded and for each $s=0(1)r$ we have $|\gamma_{rs}| \leq \max_{s=0(1)r} |\gamma_{rs}| \leq C < \infty$, then

$$\sum_{s=0}^{r} \frac{\gamma_{rs}}{n_s^{p+r}} e_{p+r}(\frac{\nu}{n}) = O[n^{-(p+r)}]$$

which brings our proof to an end.

Polynomials are not a unique class of functions for the interpolation of T_{s0}. Bulirsch and Stoer have shown ([10], [11]) that one can use the adequate classes of rational functions in the Richardson extrapolation. In this case we accept as the class $\{C_r\}_{r \in N}$ such a class of interpolating functions from [0,1] into R that each set C_r consists of rational functions of the form

$$\chi(x) = \frac{\phi(x)}{\psi(x)},$$

$\psi(x) \neq 0$ in [0,1], where a degree of the polynomial ϕ is not greater than [r/2] and a degree of ψ is not greater than r-[r/2].

After the above theoretical justifications, we now consider a practical realization of extrapolation methods. Let us denote h=1/n.

In order to determine T_{0r}, in polynomial extrapolation we use the Neville algorithm of the form (p=1 or p=2)

$$T_{s,0} := \zeta(h_s); \quad s=0(1)r;$$

$$T_{s,j} := T_{s+1,j-1} + \frac{T_{s+1,j-1} - T_{s,j-1}}{\left(\frac{h_s}{h_{s+j}}\right)^p - 1}; \qquad (1.3.55)$$

$$j=1(1)r;\ s=0(1)r-j.$$

These recursive relationships can be obtained from (1.3.53) if we have the expansion (1.3.46) with respect to the p-th power of h_s and p=1 or p=2. If p is arbitrary, but $h_s=h/k^s$, where $k>1$, then the algorithm to calculate T_{0r} has the form

$$T_{s,0} := \zeta(h_s); \quad s=0(1)r;$$

$$T_{s,j} := T_{s+1,j-1} + \frac{T_{s+1,j-1} - T_{s,j-1}}{k^{p+j-1}-1}; \qquad (1.3.56)$$

$$j=1(1)r;\ s=0(1)r-j.$$

For an arbitrary exactly increasing sequence of natural numbers $\{\beta_s\}$ and $p>2$ the adequate algorithms for the recursive computation of T_{0r} can also be obtained from (1.3.53) though in this case they have not as simple a form so (1.3.55) or (1.3.56). The formulas (1.3.55) and (1.3.56) lead to the construction of interpolating arrays shown in Fig.4a.

In the case of rational extrapolation, the quantity T_{0r} is not a linear combination of $T_{s0} \equiv \zeta(h_s)$ but it depends on T_{s0} nonlinearly. Bulirsch and Stoer have shown ([10]) that for p=1 or p=2 the quantity T_{0r} can be obtained using the following algorithm

$$T_{s,-1} := 0; \quad s=0(1)r;$$

$$T_{s,0} := \zeta(h_s); \quad s=0(1)r; \qquad (1.3.57)$$

$$T_{s,j} := T_{s+1,j-1} + \frac{T_{s+1,j-1} - T_{s,j-1}}{\left(\frac{h_s}{h_{s+j}}\right)^p \left(1 - \frac{T_{s+1,j-1} - T_{s,j-1}}{T_{s+1,j-1} - T_{s+1,j-2}}\right) - 1};$$

$$j=1(1)r;\ s=0(1)r-j.$$

Moreover, for p=2 we have

$$T_{0r} = z + O(h^{2r+2}).$$

From (1.3.57) we obtain an extrapolation array shown in Fig.4b.

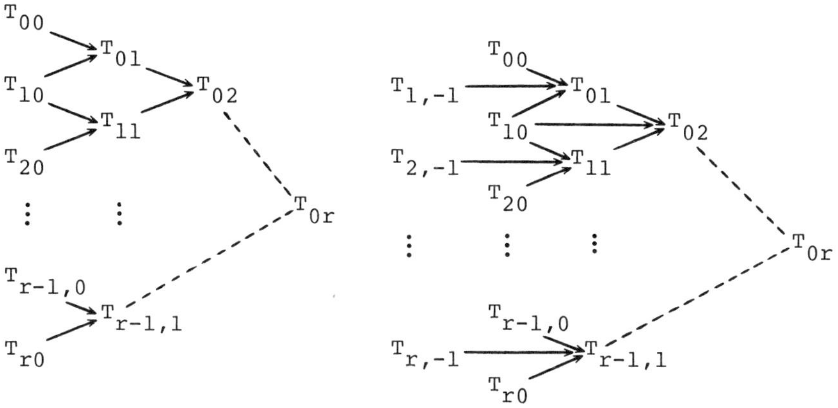

a) polynomial extrapolation   b) rational extrapolation

Fig.4. Extrapolation arrays

In extrapolation methods we use arbitrary discrete methods for which there exist adequate expansions. As we have noted above, more effective algorithms are obtained for the methods possessing asymptotic expansions with respect to the even powers of h. Moreover, from such algorithms we require an explicity of used discretization method (in order to obtain the solution quickly) and its absolute stability (see e.g.[86]).

In extrapolation algorithms the Gragg method ([35]) is used most often. This method is of the form

$$\begin{aligned} \eta^o &= z^o, \\ \eta^1 &= \eta^o+hf(x_o,\eta^o), \\ \eta^\nu &= \eta^{\nu-2}+2hf(x_{\nu-1},\eta^{\nu-1}), \quad \nu=2(1)n-1, \\ \eta^n &= \tfrac{1}{2}(\eta^{n-1}+\eta^{n-2})+\tfrac{h}{2}\, f[x_n,\eta^{n-2}+2hf(x_{n-1},\eta^{n-1})] \\ &\quad +hf(x_{n-1},\eta^{n-1}). \end{aligned} \tag{1.3.58}$$

Not taking into account the last step $\nu=n$, the above method is the two-step midpoint method (1.3.38) with the explicit starting procedure. But the midpoint method is weakly stable (see e.g.[47],[54],[69]) and it means that the main term of error, occuring in an asymptotic expansion, gives evidence of oscilation. Owing to the last step, we get rid of oscilation and we can prove

Theorem 1.9. If the function f, occuring in the initial value problem (1.1.1)-(1.1.2), is differentiable 2J times and its derivatives fulfil in some neighbourhood of the exact solution z the Lipschitz condition, then

$$\zeta(x_n,h) = z(x_n) + \sum_{j=1}^{J} h^{2j}\tilde{e}_{2j}(x_n) + O(h^{2J+1}).$$

The proof of Th.1.9 occurs e.g. in [86].

Let us note that for the midpoint method we have ([47],[86], [87])

$$\zeta(x_\nu,h) = z(x_\nu) + \sum_{j=1}^{J} h^{2j}[e_{2j,1}(x_\nu)+(-1)^\nu e_{2j,2}(x_\nu)] + O(h^{2J+1}); \quad \nu=1(1)n;$$

with the oscillating term $(-1)^\nu e_{2j,2}(x)$.

An extrapolation method can be considered as a method with variable order (succeeding columns of the array), with variable step size (taking into account the basic step size $h=1/n$) or with a variable number of stages (succeeding rows of the array). This means that to achieve the desired accuracy we can increase the order or the number of stages or decrease the step size. The increase of order causes the increase of rounding errors. From practice it is known that the Richardson extrapolation gives the best result with respect to the efficiency of the method if $r\simeq 6$. Of course, the existence of adequate asymptotic expansions up to the order $J\geq 6$ must be assumed in such a case. In a lot of papers the value $r=6$ has been accepted (see e.g.[10],[11],[29], [31]).

The choice of step size is carried out in the following way ([11]): Let h_o denote an initial step size. If the desired accuracy ε is achieved without going to sixth order extrapolation, i.e. for some $s<6$ and $1\leq k\leq s$ we have

$$|T_{k,s-k} - T_{k,s-k-1}| / \kappa < \varepsilon, \tag{1.3.59}$$

where $\kappa \simeq \max_{x\in[x_o,x_n]} |z(x)|$, then the basic step size h_o is increased to h_o' by the most recommended rule

$$h_o' = 1.5h_o.$$

If the desired accuracy is not achieved after nine stages of extrapolation, i.e.

$$| T_{4,6} - T_{3,6} |/\kappa \geq \varepsilon, \qquad (1.3.60)$$

then the basic step size must be decreased. In this purpose we halve h_o, after which the extrapolation is repeated.

Finally, if the desired accuracy is achieved after τ stages, where $6<\tau<9$, then it is recommended to modify the basic step size a little. In [35] Gragg has shown that

$$T_{\tau-6,6} - z = O(h^2_{\tau-6} \ldots h^2_\tau).$$

So, in order to obtain a sufficient accuracy of the next value T_{06}, we ought to determine h'_o from the condition

$$(h'_o \ldots h'_6)^2 \simeq (h_{\tau-6} \ldots h_\tau)^2.$$

Since for the Bulirsch sequence (1.3.51) we have

$$\frac{h_{i+1}}{h_i} = \frac{\beta_i}{\beta_{i+1}} \simeq 0.6,$$

then the new basic step size is defined by

$$h'_o = 0.9(0.6)^{\tau-6} h_o.$$

The block diagram for the rational extrapolation with the automatic step size correction described above is presented in Fig.5.

There exist some modifications of the automatic step size correction algorithm. Among others, Stoer in [88] has presented a version that leads to a significant improvement particularly in those cases where a lot of changes in the step size are required.

In problems of celestial mechanics the solution of the initial value problem on a uniform net is searched first of all. Therefore, the problem of step size choice ought to be considered in each interval $[x_\nu, x_{\nu+1}]$ ($\nu=0(1)n$; $n=[(b-a)/h]$) separately. If it appears that in realizing the above algorithm we obtain $h'_o>h$, then the order of extrapolation method must be decreased to $r=k$ (see (1.3.59)). It can also appear that after a few integration steps in the interval $[x_\nu, x_{\nu+1}]$, the next l-th step gives the solution at the point $x_{\nu_l} > x_{\nu+1}$.

In those cases we carry out the last integration step once more with the step $x_{\nu+1} - x_{\nu_{l-1}}$, remembering the value of step size calculated previously to begin an integration in the interval $[x_{\nu+1}, x_{\nu+2}]$. The first integration in [a,b] we

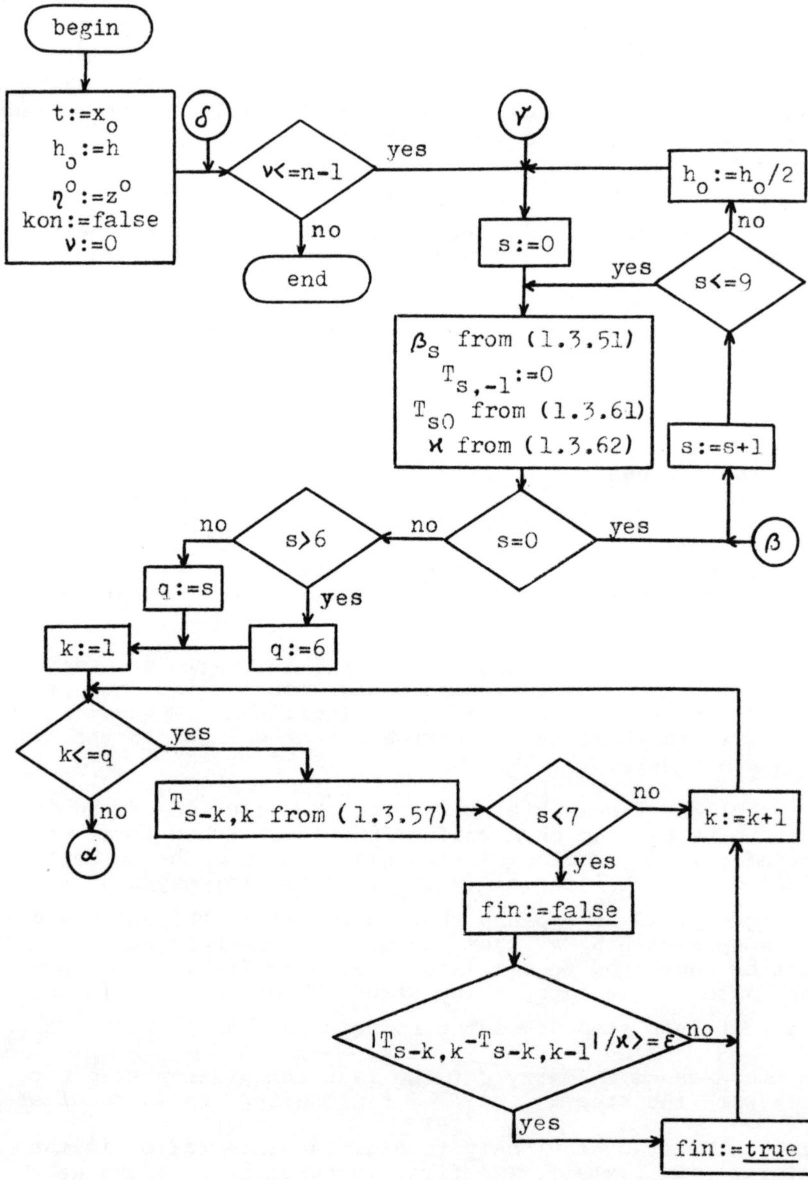
begin
t:=x$_o$
h$_o$:=h
η^o:=z^o
kon:=false
ν:=0
δ
ν<=n-1
yes
no
end
γ
h$_o$:=h$_o$/2
no
s:=0
yes
s<=9
β_s from (1.3.51)
T$_{s,-1}$:=0
T$_{s0}$ from (1.3.61)
$\varkappa$ from (1.3.62)
s:=s+1
no
s>6
no
s=0
yes
β
q:=s
yes
k:=1
q:=6
k<=q
yes
T$_{s-k,k}$ from (1.3.57)
s<7
no
k:=k+1
no
α
yes
fin:=false
$|T_{s-k,k}-T_{s-k,k-1}|/\varkappa>=\varepsilon$
no
yes
fin:=true

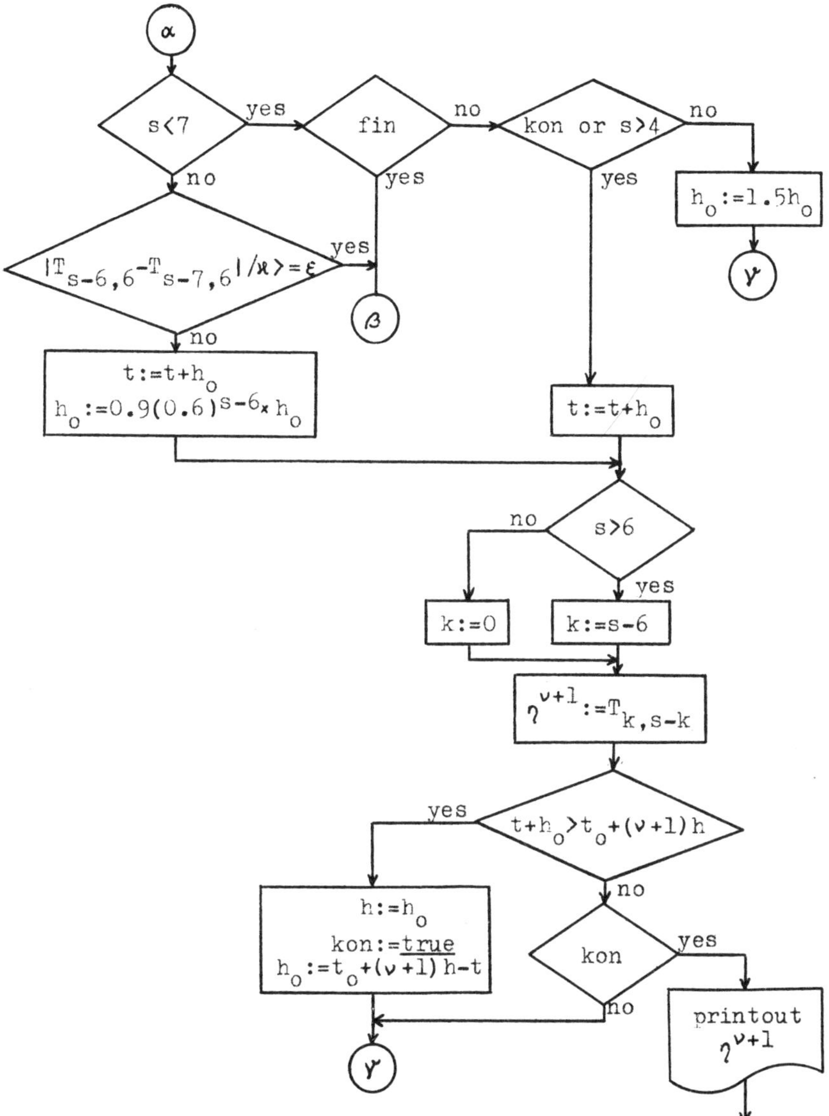

Fig.5. Block diagram for solving the initial value problem by the rational extrapolation method with the automatic step size correction

usually carry out with the step $h_o=h=(b-a)/n$. The block diagram in Fig.5 includes the remarks mentioned above.

In the interval $[x_\nu,x_{\nu+1}]$ ($\nu=0(1)n$) the Gragg method (1.3.58) has the form

$$\tilde{\eta}^o = \eta^\nu,$$
$$\tilde{\eta}^1 = \tilde{\eta}^o + \frac{h_o}{2\beta_s} f(x_\nu,\eta^\nu),$$
$$\tilde{\eta}^j = \tilde{\eta}^{j-2} + \frac{h_o}{\beta_s} f(x_{j-1},\tilde{\eta}^{j-1}); \quad j=2(1)2\beta_s-1; \qquad (1.3.61)$$
$$T_{s0} := \tilde{\eta}^{2\beta_s} = \tfrac{1}{2}(\tilde{\eta}^{2\beta_s-1}+\tilde{\eta}^{2\beta_s-2}) + \frac{h_o}{2\beta_s}\{f(x_{2\beta_s-1},\tilde{\eta}^{2\beta_s-1}) + \tfrac{1}{2}f[x_{2\beta_s},\tilde{\eta}^{2\beta_s-2}+\frac{h_o}{\beta_s} f(x_{2\beta_s-1},\tilde{\eta}^{2\beta_s-1})]\}$$
$$s=0(1)r;$$

where $x_\mu=x_\nu+\mu h_o/2\beta_s$. The quantity κ that occurs in (1.3.59) and (1.3.60) is usually determined by the formula

$$\kappa = \max_{j=0(1)2\beta_s} |\tilde{\eta}^j|. \qquad (1.3.62)$$

Apart from the papers cited previously, the extrapolation methods for solving the initial value problem are also considered in [35],[79] and [84].

1.3.4. SOME OTHER METHODS

So far we have discussed one-step, multistep and extrapolation methods. We shall now present two other classes of numerical methods for solving the initial value problem.

At first, let us present shortly the Butcher methods ([4],[14]) In these methods, like in multistep ones, we use the previous computed values of a solution at the moment t_{k-i} to obtain η^{k+1}. But at the same time in each integration step we calculate a few values in the same way as in the Runge-Kutta methods. We omit here the construction process for the Butcher methods and we only give two examples of them. For further details we refer to the references mentioned above.

The Butcher method of the sixth order has the form

$$\begin{aligned}\eta^{k+1} &= \frac{1}{31}(32\eta^{k}-\eta^{k-1})\\ &+ \frac{h}{93}(64f(x_k+h/2,\eta^{k+1/2})+12f(x_k,\eta^{k})-f(x_{k-1},\eta^{k-1})\\ &\qquad +15f\{x_{k+1},(28\eta^{k}-23\eta^{k-1})/5 \qquad (1.3.63)\\ &\qquad\qquad +h[32f(x_k+h/2,\eta^{k+1/2})-60f(x_k,\eta^{k})\\ &\qquad\qquad\quad -26f(x_{k-1},\eta^{k-1})]/15\});\\ &\qquad k=1(1)n-1;\end{aligned}$$

where

$$\eta^{k+1/2} = \eta^{k-1}+h[9f(x_k,\eta^{k})+3f(x_{k-1},\eta^{k-1})]/8.$$

From (1.3.63) it follows that in order to obtain the solution η^{k+1} we need two previously calculated solutions, namely η^{k} and η^{k-1}. It means that the above computed process can be begun after the calculation of η^{1} by an arbitrary one-step method (it is usually the Runge-Kutta method with an automatic step size correction).

The following is another example of the Butcher method.

$$\begin{aligned}\eta^{k+1} &= \frac{1}{617}(783\eta^{k}-135\eta^{k-1}-31\eta^{k-2})\\ &+ \frac{h}{3085}(2304f(x_k+h/2,\eta^{k+1/2})-135f(x_k,\eta^{k})\\ &\qquad -495f(x_k,\eta^{k-2})-39f(x_{k-2},\eta^{k-2}) \qquad (1.3.64)\\ &\qquad +465f\{x_{k+1},(540\eta^{k}-297\eta^{k-1}-212\eta^{k-2})/31\\ &\qquad\qquad + \frac{h}{155}[384f(x_k+h/2,\eta^{k+1/2})\\ &\qquad -1395f(x_k,\eta^{k})-2130f(x_{k-1},\eta^{k-1})\\ &\qquad -309f(x_{k-2},\eta^{k-2})]\}),\end{aligned}$$

where

$$\eta^{k+1/2} = (-225\eta^{k}+200\eta^{k-1}+153\eta^{k-2})/128.$$

The method (1.3.64) is of the eight order.

As follows from the above examples, the Butcher methods are multistep ones. But by comparing them with linear multistep methods (Sect.1.3.2), we see that they can be considered as the methods with an automatic step size correction because the change of step size does not imply a great increase of calculations.

Now, we take the less general class of methods in hand.

These methods refer to the solution of differential equations of the second order. Though each such equation can be reduced to the first order one (see Sect.1.1), the numerical methods for solving the differential equations of higher orders are noteworthy because in a lot of problems they are more effective.

So, let us consider some class of equations of the form

$$y'' = f(x,y). \tag{1.3.65}$$

The most popular methods for solving such equations are the Störmer methods,

$$\eta^{k+1} = 2\eta^k - \eta^{k-1} + h^2 \sum_{i=0}^{\mu} b_{\mu-i-1} f(x_{k-i+1}, \eta^{k-i+1}); \quad k=1(1)n-1. \tag{1.3.66}$$

The methods of the form (1.3.66) can be obtained requiring for the expression

$$\frac{y(x_{k+1}) - 2y(x_k) + y(x_{k-1})}{h^2} - \sum_{i=0}^{\mu} b_{\mu-i+1} y''(x_{k-i})$$

the highest possible order with respect to h.

Let us note that according to the value of the coefficient $b_{\mu+1}$, from (1.3.66) we obtain an explicit ($b_{\mu+1}=0$) or implicit ($b_{\mu+1}\neq 0$) method. These methods are usually given by the formulas

$$\eta^{k+1} = 2\eta^k - \eta^{k-1} + h^2 \sum_{p=0}^{\mu} \beta_p \nabla^p f(x_k, \eta^k),$$

$$\eta^{k+1} = 2\eta^k - \eta^{k-1} + h^2 \sum_{p=0}^{\mu} \gamma_p \nabla^p f(x_{k+1}, \eta^{k+1}),$$

where ∇ denotes the forward difference operator and

$$\{\beta_p\}_{p=0}^{\mu} = \{ 1, 0, \tfrac{1}{12}, \tfrac{19}{240}, \tfrac{3}{40}, \ldots \},$$

$$\{\gamma_p\}_{p=0}^{\mu} = \{ 1, -1, \tfrac{1}{12}, 0, -\tfrac{1}{240}, -\tfrac{1}{240}, \ldots \}.$$

The brief survey of numerical methods for solving the initial value problem, which has been given in this chapter, does not present all well-known methods in full particulars. Our purpose was only to give the reader a survey of the most popular and most often used methods. More specific methods used for solving the different variants of the N-body problem will be described in the next chapters.

Chapter 2

The General N-body Problem

2.1. EQUATIONS OF MOTION

The N-body problem consists of determining the motion of N material points attracting one another in pairs according to the Newton law of gravity on the condition that the masses of these points and their positions and velocities at some initial moment are known. This problem is applied, among others, to the study of motion of the planets of the solar system, because they can be regarded as material points considering their mutual great distances.

The general solution of the N-body problem obtained by analytical methods is unknown today. Strict solutions are known for N=2. In the case of N=3 strict solutions of practical importance are known only in some special cases. It is noteworthy that in 1912 K.F.Sundman came up with a formal solution of the general three-body problem in the form of convergent series ([90]). However, as it has been shown in the later papers of Belorizky ([6],[7],[8]), these series are most slowly convergent. Using the Sundman series, it is practically imposible to obtain the solution with sufficient accuracy even though the calculations would carry out on computers. Therefore, in modern times the numerical methods for solving the N-body problem (for $N>2$) are used.

Let us consider an isolated system of N material points attracting one another in pairs according to the Newtonian law of gravity. It means that between any two points there is a force proportional to the product of the masses of both points and inversely proportional to the second power of their distance. Let us denote the masses of material points by m_i ($i=1(1)N$), the coordinates in the inertial frame of references by x_{li} ($l=1,2,3$) and the components of velocities by v_{li}. Let $x_{li}(t_o)=x_{li}$, $v_{li}(t_o)=v^o_{li}$, where t_o denotes some initial moment. Then the motion of N material points is described by a system of differential equations of the form ([9],[22],[23],[72],[78],[81],[83],[91])

$$\dot{x}_{li} = v_{li}, \quad x_{li}(t_o) = x^o_{li}, \qquad (2.1.1)$$
$$\dot{v}_{li} = F_{li}/m_i, \quad v_{li}(t_o) = v^o_{li};$$
$$l=1,2,3; \; i=1(1)N;$$

where a dot denotes the operator d/dt and F_{li} is the l-th components of the gravitational force acting on the i-th material points, i.e.

$$F_{li} = -Gm_i \sum_{\substack{j=1\\ j\neq i}}^{N} m_j \frac{x_{li}-x_{lj}}{r^3_{ij}} . \qquad (2.1.2)$$

In the above formula G denotes the gravitational constant and r_{ij} is a distance between i-th and j-th point, i.e.

$$r_{ij} = [\sum_{l=1}^{3} (x_{li}-x_{lj})^2]^{1/2}. \qquad (2.1.3)$$

The equations of motion (2.1.1)-(2.1.2) in an inertial frame of reference are called the equations of absolute motion. The problem (2.1.1) is, of course, the initial value problem (compare (1.1.1)-(1.1.2)).

The gravitational constant $G=k^2$, where k is the Gauss constant and

$$k=0.01720209895$$

is measured in (astronomical unit)^3(mass of the Sun)$^{-1}$(efemeridal day)$^{-2}$. In the problem of motion of the planets of the solar system more often the following units are used: astronomical unit, mass of the earth, sideral year. For these units we have

$$G=0.000118568412l . \qquad (2.1.4)$$

The equations of absolute motion contain 6N variable x_{li} and v_{li} (l=1,2,3; i=1(1)N) and compose a system of 6N ordinary differential equations of the first order. In order to solve this system one should anticipate 6N constants associated with the motion, i.e. 6N functions of the x_{li}, v_{li} and t which remain constant during the motion. However, for the system (2.1.1)-(2.1.2) we know only ten ones. These functions express the well-known laws of mechanics, namely the conservation of linear and angular momenta for isolated system of material points, the uniformly and in a straight line motion of the center of mass and the conservation of total energy.

Six of the constants are easy to derive simply by adding the equations

$$m_i \dot{v}_{li} = F_{li}; \quad l=1,2,3; \; i=1(1)N.$$

Adding these equations with respect to i and taking into account (2.1.2), we have

$$\sum_{i=1}^{N} m_i \dot{v}_{li} = -G \sum_{i=1}^{N} \sum_{\substack{j=1 \\ j\neq i}}^{N} m_i m_j \frac{x_{li}-x_{lj}}{r_{ij}^3} = 0, \quad (2.1.5)$$

because for each term of the form $(x_{li}-x_{lj})/r_{ij}^3$ we can find the term $(x_{lj}-x_{li})/r_{ji}^3$ and, of course, $r_{ij}=r_{ji}$. Integrating (2.1.5), we obtain

$$\sum_{i=1}^{N} m_i v_{li} = c_l; \quad l=1,2,3; \qquad (2.1.6)$$

where c_l=const. The above equations express the constancy of linear momentum of isolated system of N material points along the length of each axis of inertial frame.

The equations (2.1.6) we can rewrite as

$$\sum_{i=1}^{N} m_i \dot{x}_{li} = c_l.$$

Hence, after integration, we obtain

$$\sum_{i=1}^{N} m_i x_{li} = c_l t + c_{l+3}; \quad l=1,2,3; \qquad (2.1.7)$$

where c_{l+3}=const. Thus, we get the next three constants. Determining coordinates of the center of mass of N material points by the formulas

$$x_{lb} = \frac{1}{M} \sum_{i=1}^{N} m_i x_{li}; \quad l=1,2,3; \qquad (2.1.8)$$

where

$$M = \sum_{i=1}^{N} m_i,$$

we can rewrite the equation (2.1.7) in the form

$$Mx_{lb} = c_l t + c_{l+3}; \quad l=1,2,3. \qquad (2.1.9)$$

The equations (2.1.8) and (2.1.6) yield

$$\dot{x}_{lb} = \frac{1}{M} \sum_{i=1}^{N} m_i \dot{x}_{li} = c_l/M,$$

and hence

$$v_b = \left(\sum_{l=1}^{3} \dot{x}_{lb}^2 \right)^{1/2} = \frac{1}{M} \left(\sum_{l=1}^{3} c_l^2 \right)^{1/2}. \qquad (2.1.10)$$

From (2.1.9) and (2.1.10) it follows that the center of mass moves in a straight line with uniform velocity.

Now, let us compose the expressions

$$m_i(x_{pi}\dot{v}_{qi} - x_{qi}\dot{v}_{pi}),$$

where $(p,q) \in \{(2,3),(3,1),(1,2)\}$. From (2.1.1)-(2.1.2) we obtain

$$\begin{aligned} m_i(x_{pi}\dot{v}_{qi} - x_{qi}\dot{v}_{pi}) &= -Gm_i \sum_{\substack{j=1 \\ j\neq i}}^{N} m_j \frac{x_{pi}(x_{qi}-x_{qj})}{r_{ij}^3} \\ &\quad +Gm_i \sum_{\substack{j=1 \\ j\neq i}}^{N} m_j \frac{x_{qi}(x_{pi}-x_{pj})}{r_{ij}^3} \\ &= -Gm_i \sum_{\substack{j=1 \\ j\neq i}}^{N} \frac{m_j}{r_{ij}^3} [x_{pi}(x_{qi}-x_{qj}) - x_{qi}(x_{pi}-x_{pj})] \\ &= -Gm_i \sum_{\substack{j=1 \\ j\neq i}}^{N} \frac{m_j}{r_{ij}^3} (x_{pi}x_{qj} - x_{qi}x_{pj}). \end{aligned}$$

Hence

$$\sum_{i=1}^{N} m_i(x_{pi}\dot{v}_{qi} - x_{qi}\dot{v}_{pi}) = 0. \qquad (2.1.11)$$

Integration of (2.1.11) yields the next three constants of motion, namely we have

$$\sum_{i=1}^{N} m_i(x_{pi}v_{qi} - x_{qi}v_{pi}) = c_{l+6}, \qquad (2.1.12)$$

where $l=1$ for $(p,q)=(2,3)$, $l=2$ for $(p,q)=(3,1)$ and $l=3$ for $(p,q)=(1,2)$. The equations (2.1.12) express the constancy of angular momentum with respect to each axis.

In order to determine the last, tenth constant of motion, let us introduce the function $V=V(x_1,x_2,x_3)$ such that

$$V = \frac{G}{2} \sum_{i=1}^{N} \sum_{\substack{j=1 \\ j\neq i}}^{N} \frac{m_i m_j}{r_{ij}} . \qquad (2.1.13)$$

The function V is called an attracting potential for isolated system of N material points. Taking into account (2.1.13), the equation of motion (2.1.1)-(2.1.2) we can write as follows

$$\begin{aligned} \dot{x}_{li} &= v_{li}, \\ m_i \dot{v}_{li} &= \partial V/\partial x_{li}; \quad l=1,2,3; \; i=1(1)N. \end{aligned} \qquad (2.1.14)$$

Let us multiply the second equation of (2.1.14) by v_{1i}, v_{2i} and v_{3i} in turn and then sum up the obtained expressions and add them up with respect to i. We obtain

$$\sum_{i=1}^{N} m_i \sum_{l=1}^{3} v_{li}\dot{v}_{li} = \sum_{i=1}^{N} \sum_{l=1}^{3} \frac{\partial V}{\partial x_{li}} v_{li} . \qquad (2.1.15)$$

Let us note that

$$\frac{dV}{dt} = \sum_{i=1}^{N} \sum_{l=1}^{3} \frac{\partial V}{\partial x_{li}} \frac{dx_{li}}{dt} = \sum_{i=1}^{N} \sum_{l=1}^{3} \frac{\partial V}{\partial x_{li}} v_{li} .$$

So, integrating the equation (2.1.15), we get

$$\frac{1}{2} \sum_{i=1}^{N} m_i \sum_{l=1}^{3} v_{li}^2 = V + c_{10},$$

where c_{10}=const. Denoting by K the left-hand side of this equation, we have

$$K - V = c_{10} . \qquad (2.1.16)$$

The equation (2.1.16) expresses the principle of energy for the isolated system of N material points. This energy consists of the sum of kinetic and potential energy, i.e.

$$E = \frac{1}{2} \sum_{i=1}^{N} m_i \left(\sum_{l=1}^{3} v_{li}^2 - G \sum_{\substack{j=1 \\ j\neq i}}^{N} \frac{m_j}{r_{ij}} \right). \qquad (2.1.17)$$

The obtained ten constants are all known constants for the N-body problem considered in inertial rectangular frame of references. Since it is impossible to determine any fixed point in the space, we can not determine the first six constants of motion. However, as it has been shown, the center of mass moves in a straight line with uniform velocity. Transforming the origin of a coordinate system to this center, we obtain an inertial frame. It can be shown that in the inertial frame with the origin in the center of mass the equations of motion (2.1.1)-(2.1.2) have the same form, but in this case they are the system of the order 6N-6.

We now present an example of an analytical solution of the N-body problem which we will use in the next sections to verify numerical methods for solving this problem. As it has been mentioned, it is impossible to solve immediately the equations of absolute motion (2.1.1)-(2.1.2) in an arbitrary inertial frame even though N=2. We still need 6N-10=2 constants of motion. These constants can be determined if we impose some additional conditions on an absolute motion of two material points. Of course, another way to solve such equations is to put the origin of frame in the center of mass.

As is well-known ([9],[72],[83],[91]), the motion of an isolated system of two material points is a plane motion. So, let us take the axes of an inertial rectangular frame in such a way that the plane $x_3=0$ is a plane of motion. Moreover, let us assume that the material points P_1 and P_2 with the masses m_1 and m_2 respectively are in a constant mutual distance, i.e.

$$r = \left[\sum_{l=1}^{2} (x_{l1}-x_{l2})^2 \right]^{1/2} = a = \text{const.} \qquad (2.1.18)$$

This condition permits us to determine the needed constants. Differentiation of (2.1.18) yields

$$\sum_{l=1}^{2} (x_{l1}-x_{l2})(v_{l1}-v_{l2}) = 0. \qquad (2.1.19)$$

Now, let $x_{11}^o=a$, $v_{21}^o=2\pi a$, $x_{li}^o=v_{li}^o=0$ for other l and i at the outset $t_o=0$. From (2.1.6), (2.1.7), (2.1.12), (2.1.16), (2.1.18) and (2.1.19) we obtain

$$v_{12} = -\frac{m_1}{m_2} v_{11},$$

$$v_{22} = \frac{m_1}{m_2}(2\pi a-v_{21}),$$

$$x_{12} = \frac{m_1}{m_2}(a-x_{11}),$$

$$x_{22} = \frac{m_1}{m_2}(2\pi at-x_{21}), \qquad (2.1.20)$$

$$m_1(x_{11}v_{21}-x_{21}v_{11})+m_2(x_{12}v_{22}-x_{22}v_{12}) = 2\pi a^2 m_1,$$

$$\frac{1}{2}[m_1(v_{11}^2+v_{21}^2)+m_2(v_{12}^2+v_{22}^2)] = 2\pi^2a^2m_1,$$
$$(x_{11}-x_{12})(v_{11}-v_{12})+(x_{21}-x_{22})(v_{21}-v_{22}) = 0,$$
$$(x_{11}-x_{12})^2+(x_{21}-x_{22})^2 = a^2.$$

This system consists of eight equations with the same number of unknowns. Taking into account the first four equations of (2.1.20), we can reduce the system to the form

$$\begin{aligned}
&(m_1+m_2)(x_{11}v_{21}-x_{21}v_{11})-am_1v_{21}-2\pi am_1x_{11}+2\pi am_1tv_{11} \\
&\qquad = 2\pi a^2(m_2-m_1), \\
&(m_1+m_2)(v_{11}^2+v_{21}^2)-4\pi am_1v_{21} = 4\pi^2a^2(m_2-m_1), \\
&(m_1+m_2)^2(x_{11}v_{11}+x_{21}v_{21})-m_1(m_1+m_2)(v_{11}+2\pi x_{21}+2\pi tv_{21})a \\
&\qquad = -4\pi^2a^2tm_1^2, \\
&(m_1+m_2)^2(x_{11}^2+x_{21}^2)-2m_1(m_1+m_2)(x_{11}+2\pi tx_{21})a \\
&\qquad = a^2(m_2^2-m_1^2)-4\pi^2a^2t^2m_1^2.
\end{aligned} \tag{2.1.21}$$

Let us rewrite (2.1.21) as follows

$$(x_{11}-\mu)(v_{21}-2\pi\mu)-(x_{21}-2\pi t\mu)v_{11} = 2\pi(a-\mu)^2,$$
$$v_{11}^2+(v_{21}-2\pi\mu)^2 = 4\pi^2(a-\mu)^2,$$
$$(x_{11}-\mu)v_{11}+(x_{21}-2\pi t\mu)(v_{21}-2\pi\mu) = 0,$$
$$(x_{11}-\mu)^2+(x_{21}-2\pi t\mu)^2 = (a-\mu)^2,$$

where $\mu=am_1/(m_1+m_2)$. The second and fourth equation present circles (while t is fixed) which can be described in a parametrical form as

$$x_{11} = \mu+(a-\mu)\cos\alpha,$$
$$x_{21} = 2\pi t\mu+(a-\mu)\sin\alpha,$$
$$v_{11} = 2\pi(a-\mu)\cos\beta,$$
$$v_{21} = 2\pi\mu+2\pi(a-\mu)\sin\beta.$$

From the remaining two equations we see immediately that $\beta=\alpha+\pi/2$. It is also easy to show that $\alpha=2\pi t$. Finally, we get the solution of the form

$$x_{11} = \frac{a}{m_1+m_2}(m_1+m_2\cos 2\pi t),$$

$$x_{21} = \frac{a}{m_1+m_2}(2\pi t m_1+m_2\sin 2\pi t),$$
$$v_{11} = -\frac{2a\pi m_2}{m_1+m_2}\sin 2\pi t, \tag{2.1.22}$$
$$v_{21} = \frac{2a}{m_1+m_2}(m_1+m_2\cos 2\pi t).$$

Substituting (2.1.22) in the first four equations of (2.1.20), we have

$$x_{12} = \frac{am_1}{m_1+m_2}(1-\cos 2\pi t),$$
$$x_{22} = \frac{am_1}{m_1+m_2}(2\pi t-\sin 2\pi t),$$
$$v_{12} = \frac{2a\pi m_1}{m_1+m_2}\sin 2\pi t, \tag{2.1.23}$$
$$v_{22} = \frac{2a\pi m_1}{m_1+m_2}(1-\cos 2\pi t).$$

For $a=1$, $m_1=1$, $m_2=332958$ and $t\in[0,1]$ we obtain the solution which at the moments $t_\nu=\nu/100$ for $\nu=20(20)100$ is presented in Table V. The results from this table will be used in the next sections to verify numerical methods for solving the problem (2.1.1)-(2.1.2).

2.2. THE APPLICATION OF CONVENTIONAL NUMERICAL METHODS

In view of a great number of different methods used for solving the problem (2.1.1)-(2.1.2), it is not possible to analyze all of them in this book. Therefore, we restrict ourselves to a few methods described in Ch.1. In practice some of them are not used, but it seems to be advisable to illustrate this fact by examples. In our analysis we will pay special attention to the quantities (i)-(v) mentioned at the end of Sect.1.2 and to the constants of motion.

Let us begin our consideration by formalizing the problem. Let us assume that in an inertial frame of reference at the outset the coordinates x^o_{li} and the components of velocities v^o_{li} ($l=1,2,3$; $l=1(1)N$) are given. Let m_i denote a mass

Table V. Exact solution of the two-body problem in inertial frame

ν	i	l	x_{li}	v_{li}	c_p
20	1	1	0.309019069655	-5.975646382335	c_1=0.000000000000
		2	0.951057434067	1.941624078091	
	2	1	0.000002075280	0.000017947148	c_2=6.283185307180
		2	0.000000917772	0.000013039366	
40	1	1	-0.809011561223	-3.693152569034	c_4=1.000000000000
		2	0.587791035253	-5.083169554812	
	2	1	0.000005433152	0.000011091947	c_5=0.000000000000
		2	0.000005782961	0.000034137503	
60	1	1	-0.809011561223	3.693152569034	c_9=6.283185307180
		2	-0.587772164506	-5.083169554812	
	2	1	0.000005433152	-0.000011091947	c_{10}=
		2	0.000013087787	0.000034137503	-19.73909255381
80	1	1	0.309019069655	5.975646382335	
		2	-0.951038563320	1.941624078092	
	2	1	0.000002075280	-0.000017947148	
		2	0.000013039366	0.000018077771	
100	1	1	1.000000000000	0.000000000000	
		2	0.000018870748	6.283185307180	
	2	1	0.000000000000	0.000000000000	
		2	0.000018870748	0.000000000000	

ν - moment
i - number of body
l - axis
c_p - constant of motion

of i-th body and let the motion be described by the system of differential equations (2.1.1)-(2.1.2). The solution of this system ought to be found on the interval $[t_o, t_n]$ at the points $t_\nu = t_o + \nu\Delta t$, where $\Delta t = (t_n - t_o)/n$ $(\nu = 1(1)n)$.

<u>Algorithm 1.</u> The Euler modified method with automatic step size correction

1° $H := \Delta t$;
   $S :=$ <u>true</u>;
   $k := 0$;
2° $t := t_o + k\Delta t$;
3° for l=1,2,3 and i=1(1)N:

$$x_{li}^{k+1} := x_{li}^{k} + H\left[v_{li}^{k} - \frac{HG}{2}\sum_{\substack{j=1\\ j\neq i}}^{N} m_j \frac{x_{li}^{k} - x_{lj}^{k}}{(r_{ij}^{k})^3}\right],$$

where $r_{ij}^{k} = \left[\sum_{l=1}^{3} (x_{li}^{k} - x_{lj}^{k})^2\right]^{1/2}$;

$$v_{li}^{k+1} := v_{li}^{k} - HG \sum_{\substack{j=1 \\ j\neq i}}^{N} m_j \frac{x_{li}^{k} - x_{lj}^{k} + \frac{H}{2}(v_{li}^{k} - v_{lj}^{k})}{(\tilde{r}_{ij}^{k})^3},$$

where $\tilde{r}_{ij}^{k} = \{\sum_{l=1}^{3} [x_{li}^{k} - x_{lj}^{k} + \frac{H}{2}(v_{li}^{k} - v_{lj}^{k})]^2\}^{1/2}$;

4° if S=<u>true</u> then go to 5°;
 $H:=H_1$;
 S:=<u>true</u>;
 printout x_{li}^{k+1} and v_{li}^{k+1} ($l=1,2,3$; $i=1(1)N$);
 $k:=k+1$;
 if $k\leq n-1$ then go to 2°;
 stop;

5° F:=<u>true</u>;
 for $l=1,2,3$ and $i=1(1)N$:
  $y_{li} := x_{li}^{k}$;
  $w_{li} := v_{li}^{k}$;

6° for $l=1,2,3$ and $i=1(1)N$:

$$z_{li} := y_{li} + \frac{H}{2}(w_{li} - \frac{HG}{4} \sum_{\substack{j=1 \\ j\neq i}}^{N} m_j \frac{y_{li} - y_{lj}}{r_{ij}^3}),$$

where $r_{ij} = [\sum_{l=1}^{3} (y_{li} - y_{lj})^2]^{1/2}$;

$$u_{li} := w_{li} - \frac{HG}{2} \sum_{\substack{j=1 \\ j\neq i}}^{N} m_j \frac{y_{li} - y_{lj} + \frac{H}{4}(w_{li} - w_{lj})}{\tilde{r}_{ij}^3},$$

where $\tilde{r}_{ij} = \{\sum_{l=1}^{3} [y_{li} - y_{lj} + \frac{H}{4}(w_{li} - w_{lj})]^2\}^{1/2}$;

7° if F=<u>false</u> then go to 8°;
 F:=<u>false</u>;
 for $l=1,2,3$ and $i=1(1)N$:
  $y_{li} := z_{li}$;
  $w_{li} := u_{li}$;
 go to 6°;

8° $\tilde{h} := H/[\frac{4}{3} \max_{\substack{l=1,2,3 \\ i=1(1)N}} (|x_{li}^{k+1} - z_{li}|, |v_{li}^{k+1} - u_{li}|)/\varepsilon]^{1/3}$ (ε denotes a discretization error given beforehand);
 if $\tilde{h}\leq H/3$ then:
  $H:=2\tilde{h}$;
  go to 3°;

if $\tilde{h} \geq H$ then:

if $t+2H \leq t_o+(k+1)\Delta t$ then:

$H:=2H;$

go to 3°;

$H:=t_o+(k+1)\Delta t-t;$

$H_1:=H;$

S:=<u>false</u>;

go to 3°;

for $l=1,2,3$ and $i=1(1)N$:

$x_{li}^k:=z_{li};$

$v_{li}^k:=u_{li};$

$t:=t+H;$

if $t+2\tilde{h}<t_o+(k+1)\Delta t$ then:

$H:=2\tilde{h};$

go to 3°;

$H:=t_o+(k+1)\Delta t-t;$

$H_1:=2\tilde{h};$

S:=<u>false</u>;

go to 3°.

<u>Algorithm 2.</u> The Runge-Kutta method of fourth order with automatic step size correction

$1^o,2^o,4^o,5^o,7^o$ - as in Algorithm 1;

3° for $l=1,2,3$ and $i=1(1)N$:

$$A_{li}^{(1)}:=-G\sum_{\substack{j=1\\j\neq i}}^{N} m_j\frac{x_{li}^k-x_{lj}^k}{(r_{ij}^k)^3}\text{ , where } r_{ij}^k=[\sum_{l=1}^{3}(x_{li}^k-x_{lj}^k)^2]^{1/2};$$

$$A_{li}^{(2)}:=-G\sum_{\substack{j=1\\j\neq i}}^{N} m_j\frac{x_{li}^k-x_{lj}^k+H(v_{li}^k-v_{lj}^k)/2}{(\bar{r}_{ij}^k)^3}\text{ ,}$$

$$\text{where } \bar{r}_{ij}^k=\{\sum_{l=1}^{3}[x_{li}^k-x_{lj}^k+H(v_{li}^k-v_{lj}^k)/2]^2\}^{1/2};$$

for $l=1,2,3$ and $i=1(1)N$:

$$A^{(3)}:=-G\sum_{\substack{j=1\\j\neq i}}^{N} m_j\frac{x_{li}^k-x_{lj}^k+H(v_{li}^k-v_{lj}^k)/2+H^2(A_{li}^{(1)}-A_{lj}^{(1)})/4}{(\tilde{r}_{ij}^k)^3}\text{ ,}$$

$$\text{where } \tilde{r}_{ij}^k=\{\sum_{l=1}^{3}[x_{li}^k-x_{lj}^k+H(v_{li}^k-v_{lj}^k)/2 +H^2(A_{li}^{(1)}-A_{lj}^{(1)})/4]^2\}^{1/2};$$

$$A^{(4)} := -G \sum_{\substack{j=1\\ j\neq i}}^{N} m_j \frac{x_{li}^k - x_{lj}^k + H(v_{li}^k - v_{lj}^k) + H^2(A_{li}^{(2)} - A_{lj}^{(2)})/2}{(\hat{r}_{ij}^k)^3},$$

where $\hat{r}_{ij}^k = \{\sum_{l=1}^{3} [x_{li}^k - x_{lj}^k + H(v_{li}^k - v_{lj}^k) + H^2(A_{li}^{(2)} - A_{lj}^{(2)})/2]^2\}^{1/2}$;

$$x_{li}^{k+1} := x_{li}^k + H[v_{li}^k + \frac{H}{6}(A_{li}^{(1)} + A_{li}^{(2)} + A^{(3)})];$$

$$v_{li}^{k+1} := v_{li}^k + \frac{H}{6}(A_{li}^{(1)} + 2A_{li}^{(2)} + 2A^{(3)} + A^{(4)});$$

6° for $l=1,2,3$ and $i=1(1)N$:

$$A_{li}^{(1)} := -G \sum_{\substack{j=1\\ j\neq i}}^{N} m_j \frac{y_{li} - y_{lj}}{r_{ij}^3}, \text{ where } r_{ij} = [\sum_{l=1}^{3} (y_{li} - y_{lj})^2]^{1/2};$$

$$A_{li}^{(2)} := -G \sum_{\substack{j=1\\ j\neq i}}^{N} m_j \frac{y_{li} - y_{lj} + H(w_{li} - w_{lj})/4}{\bar{r}_{ij}^3},$$

where $\bar{r}_{ij} = \{\sum_{l=1}^{3} [y_{li} - y_{lj} + H(w_{li} - w_{lj})/4]^2\}^{1/2}$;

for $l=1,2,3$ and $i=1(1)N$:

$$A^{(3)} := -G \sum_{\substack{j=1\\ j\neq i}}^{N} m_j \frac{y_{li} - y_{lj} + H(w_{li} - w_{lj})/4 + H^2(A_{li}^{(1)} - A_{lj}^{(1)})/16}{\tilde{r}_{ij}^3},$$

where $\tilde{r}_{ij} = \{\sum_{l=1}^{3} [y_{li} - y_{lj} + H(w_{li} - w_{lj})/4 + H^2(A_{li}^{(1)} - A_{lj}^{(1)})/16]^2\}^{1/2}$;

$$A^{(4)} := -G \sum_{\substack{j=1\\ j\neq i}}^{N} m_j \frac{y_{li} - y_{lj} + H(w_{li} - w_{lj})/2 + H^2(A_{li}^{(2)} - A_{lj}^{(2)})/8}{\hat{r}_{ij}^3},$$

where $\hat{r}_{ij} = \{\sum_{l=1}^{3} [y_{li} - y_{lj} + H(w_{li} - w_{lj})/2 + H^2(A_{li}^{(2)} - A_{lj}^{(2)})/8]^2\}^{1/2}$;

$$z_{li} := y_{li} + H[w_{li} + \frac{H}{6}(A_{li}^{(1)} + A_{li}^{(2)} + A^{(3)})];$$

$$u_{li} := w_{li} + \frac{H}{6}(A_{li}^{(1)} + 2A_{li}^{(2)} + 2A^{(3)} + A^{(4)});$$

8° $h := H/[\frac{16}{15} \max_{\substack{l=1,2,3\\ i=1(1)N}} (|x_{li}^{k+1} - z_{li}|, |v_{li}^{k+1} - u_{li}|)/\varepsilon]^{1/5}$;

continue as in Algorithm 1, 8°.

Algorithm 3. **The Adams-Bashforth method of fourth order**

1° compute x_{li}^{k} and v_{li}^{k} for k=1,2,3 (l=1,2,3; i=1(1)N) by the Runge-Kutta method of fourth order with automatic step size correction (see Algorithm 2);

2° k:=3;

3° for l=1,2,3 and i=1(1)N:

$$x_{li}^{k+1}:=x_{li}^{k}+\frac{\Delta t}{2.4}(5.5v_{li}^{k}-5.9v_{li}^{k-1}+3.7v_{li}^{k-2}-0.9v_{li}^{k-3});$$

$$v_{li}^{k+1}:=v_{li}^{k}-\frac{\Delta tG}{2.4}\sum_{\substack{j=1\\ j\neq i}}^{N} m_j[5.5\frac{x_{li}^{k}-x_{lj}^{k}}{(r_{ij}^{k})^3}-5.9\frac{x_{li}^{k-1}-x_{lj}^{k-1}}{(r_{ij}^{k-1})^3}+3.7\frac{x_{li}^{k-2}-x_{lj}^{k-2}}{(r_{ij}^{k-2})^3}-0.9\frac{x_{li}^{k-3}-x_{lj}^{k-3}}{(r_{ij}^{k-3})^3}],$$

where $r_{ij}^{p}=[\sum_{l=1}^{3}(x_{li}^{p}-x_{lj}^{p})^2]^{1/2}$ (p=k-3(1)k);

printout x_{li}^{k+1}, v_{li}^{k+1} (l=1,2,3; i=1(1)N);

4° k:=k+1;
if k<n then go to 3°;
stop.

Algorithm 4. **Seventh order method of Adams-Bashforth**

1° compute x_{li}^{k} and v_{li}^{k} for k=1(1)6 (l=1,2,3; i=1(1)N) by the Runge-Kutta method of fourth order with automatic step size correction (see Algorithm 2);

2° k:=6;

3° for l=1,2,3 and i=1(1)N:

$$x_{li}^{k+1}:=x_{li}^{k}+\frac{\Delta t}{0.60480}(1.98721v_{li}^{k}-4.47288v_{li}^{k-1}+7.05549v_{li}^{k-2}-6.88256v_{li}^{k-3}+4.07139v_{li}^{k-4}-1.34472v_{li}^{k-5}+0.19087v_{li}^{k-6});$$

$$v_{li}^{k+1}:=v_{li}^{k}-\frac{\Delta tG}{0.60480}\sum_{\substack{j=1\\ j\neq i}}^{N} m_j[1.98721\frac{x_{li}^{k}-x_{lj}^{k}}{(r_{ij}^{k})^3}-4.47288\frac{x_{li}^{k-1}-x_{lj}^{k-1}}{(r_{ij}^{k-1})^3}$$

$$+7.05549\frac{x_{li}^{k-2}-x_{lj}^{k-2}}{(r_{ij}^{k-2})^3}-6.88256\frac{x_{li}^{k-3}-x_{lj}^{k-3}}{(r_{ij}^{k-3})^3}$$

$$+4.07139\,\frac{x_{li}^{k-4}-x_{lj}^{k-4}}{(r_{ij}^{k-4})^3}-1.34472\,\frac{x_{li}^{k-5}-x_{lj}^{k-5}}{(r_{ij}^{k-5})^3}$$

$$+0.19087\,\frac{x_{li}^{k-6}-x_{lj}^{k-6}}{(r_{ij}^{k-6})^3}],$$

where $r_{ij}^p=[\sum_{l=1}^{3}(x_{li}^p-x_{lj}^p)^2]^{1/2}$ $(p=k-6(1)k)$;

4° as in Algorithm 3.

<u>Algorithm 5.</u> Seventh order predictor-corrector method based on the Adams formulas

1° compute x_{li}^k and v_{li}^k for k=1,2,3,4,5 (l=1,2,3; i=1(1)N) by the Runge-Kutta method of fourth order with automatic step size correction (see Algorithm 2);

2° k:=5;

3° for l=1,2,3 and i=1(1)N:

$$x_{li}^{k+1}:=x_{li}^k+\frac{\Delta t}{1.440}(4.227v_{li}^k-7.673v_{li}^{k-1}+9.482v_{li}^{k-2}-6.798v_{li}^{k-3}$$
$$+2.627v_{li}^{k-4}-0.425v_{li}^{k-5});$$

$$v_{li}^{k+1}:=v_{li}^k-\frac{\Delta tG}{1.440}\sum_{\substack{j=1\\j\neq i}}^{N}m_j[4.227\,\frac{x_{li}^k-x_{lj}^k}{(r_{ij}^k)^3}-7.673\,\frac{x_{li}^{k-1}-x_{lj}^{k-1}}{(r_{ij}^{k-1})^3}$$
$$+9.482\,\frac{x_{li}^{k-2}-x_{lj}^{k-2}}{(r_{ij}^{k-2})^3}-6.798\,\frac{x_{li}^{k-3}-x_{lj}^{k-3}}{(r_{ij}^{k-3})^3}$$
$$+2.627\,\frac{x_{li}^{k-4}-x_{lj}^{k-4}}{(r_{ij}^{k-4})^3}-0.425\,\frac{x_{li}^{k-5}-x_{lj}^{k-5}}{(r_{ij}^{k-5})^3}],$$

where $r_{ij}^p=[\sum_{l=1}^{3}(x_{li}^p-x_{lj}^p)^2]^{1/2}$ $(p=k-5(1)k)$;

4° for l=1,2,3 and i=1(1)N:

$\bar{x}_{li}^{k+1}:=x_{li}^{k+1}$;

$\bar{v}_{li}^{k+1}:=v_{li}^{k+1}$;

$$w_{li}:=6.5112v_{li}^k-4.6461v_{li}^{k-1}+3.7504v_{li}^{k-2}-2.0211v_{li}^{k-3}$$
$$+0.6312v_{li}^{k-4}-0.0863v_{li}^{k-5};$$

$$z_{li}:=\sum_{\substack{j=1\\j\neq i}}^{N}m_j[6.5112\,\frac{x_{li}^k-x_{lj}^k}{(r_{ij}^k)^3}-4.6461\,\frac{x_{li}^{k-1}-x_{lj}^{k-1}}{(r_{ij}^{k-1})^3}$$

$$+3.7504\,\frac{x_{li}^{k-2}-x_{lj}^{k-2}}{(r_{ij}^{k-2})^3}-2.0211\,\frac{x_{li}^{k-3}-x_{lj}^{k-3}}{(r_{ij}^{k-3})^3}$$

$$+0.6312\,\frac{x_{li}^{k-4}-x_{lj}^{k-4}}{(r_{ij}^{k-4})^3}-0.0863\,\frac{x_{li}^{k-5}-x_{lj}^{k-5}}{(r_{ij}^{k-5})^3}],$$

where $r_{ij}^p=[\sum_{l=1}^{3}(x_{li}^p-x_{lj}^p)^2]^{1/2}$ $(p=k-5(1)k)$;

5° for $l=1,2,3$ and $i=1(1)N$:

$$x_{li}^{k+1}:=x_{li}^k+\frac{\Delta t}{6.0480}(1.9087v_{li}^{k+1}+w_{li});$$

6° for $l=1,2,3$ and $i=1(1)N$:

$$v_{li}^{k+1}:=v_{li}^k-\frac{\Delta tG}{6.0480}[1.9087\sum_{\substack{j=1\\j\neq i}}^{N}m_j\frac{x_{li}^{k+1}-x_{lj}^{k+1}}{(r_{ij}^{k+1})^3}+z_{li}],$$

where $r_{ij}^{k+1}=[\sum_{l=1}^{3}(x_{li}^{k+1}-x_{lj}^{k+1})^2]^{1/2}$;

7° if there exist i $(i=1(1)N)$ or l $(l=1,2,3)$ such that $|\bar{x}_{li}^{k+1}-x_{li}^{k+1}|\geq\varepsilon$ or $|\bar{v}_{li}^{k+1}-v_{li}^{k+1}|\geq\varepsilon$ then:

  for $l=1,2,3$ and $i=1(1)N$:

    $\bar{x}_{li}^{k+1}:=x_{li}^{k+1}$;

    $\bar{v}_{li}^{k+1}:=v_{li}^{k+1}$;

  go to 5°;

printout x_{li}^{k+1} and v_{li}^{k+1} $(l=1,2,3;\ i=1(1)N)$;

$k:=k+1$;

if $k<n$ then go to 3°;

stop.

Algorithm 6. The Butcher method of sixth order

1° compute x_{li}^1 and v_{li}^1 $(l=1,2,3;\ i=1(1)N)$ by the Runge-Kutta method of fourth order with automatic step size correction (see Algorithm 2);

2° $k:=1$;

3° for $l=1,2,3$ and $i=1(1)N$:

$$w_{li}:=v_{li}^{k-1}-\frac{\Delta tG}{8}\sum_{\substack{j=1\\j\neq i}}^{N}m_j[9\,\frac{x_{li}^k-x_{lj}^k}{(r_{ij}^k)^3}+3\,\frac{x_{li}^{k-1}-x_{lj}^{k-1}}{(r_{ij}^{k-1})^3}],$$

where $r_{ij}^p=[\sum_{l=1}^{3}(x_{li}^p-x_{lj}^p)^2]^{1/2}$ $(p=k-1,k)$;

$z_{li}:=x_{li}^{k-1}+\frac{\Delta t}{8}(9v_{li}^{k}+3v_{li}^{k-1})$;

4° for l=1,2,3 and i=1(1)N:

$\tilde{z}_{li}:=5.6x_{li}^{k}-4.6x_{li}^{k-1}+\frac{\Delta t}{1.5}(3.2w_{li}-6.0v_{li}^{k}-2.6v_{li}^{k-1})$;

5° for l=1,2,3 and i=1(1)N:

$$x_{li}^{k+1}:=\frac{1}{3.1}(3.2x_{li}^{k}-0.1x_{li}^{k-1})+\frac{\Delta t}{9.3}\{6.4w_{li}+9.6v_{li}^{k}-7.0v_{li}^{k-1}-\Delta tG\sum_{\substack{j=1\\j\neq i}}^{N}m_j[3.2\frac{z_{li}-z_{lj}}{(\tilde{r}_{ij})^3}-6.0\frac{x_{li}^{k}-x_{lj}^{k}}{(r_{ij}^{k})^3}-2.6\frac{x_{li}^{k-1}-x_{lj}^{k-1}}{(r_{ij}^{k-1})^3}]\};$$

$$v_{li}^{k+1}:=\frac{1}{3.1}(3.2v_{li}^{k}-0.1v_{li}^{k-1})-\frac{\Delta tG}{9.3}\sum_{\substack{j=1\\j\neq i}}^{N}m_j[6.4\frac{z_{li}-z_{lj}}{(\tilde{r}_{ij})^3}+1.2\frac{x_{li}^{k}-x_{lj}^{k}}{(r_{ij}^{k})^3}-0.1\frac{x_{li}^{k-1}-x_{lj}^{k-1}}{(r_{ij}^{k-1})^3}+1.5\frac{\tilde{z}_{li}-\tilde{z}_{lj}}{(\hat{r}_{ij})^3}],$$

where $r_{ij}^{p}=[\sum_{l=1}^{3}(x_{li}^{p}-x_{lj}^{p})^2]^{1/2}$ (p=k-1,k),

$\tilde{r}_{ij}=[\sum_{l=1}^{3}(z_{li}-z_{lj})^2]^{1/2}$,

$\hat{r}_{ij}=[\sum_{l=1}^{3}(\tilde{z}_{li}-\tilde{z}_{lj})^2]^{1/2}$;

6° printout x_{li}^{k+1} and v_{li}^{k+1} (l=1,2,3; i=1(1)N);

k=k+1;

if k<n then go to 3°;

stop.

<u>Algorithm 7.</u> Polynomial extrapolation with automatic step size correction

1° $t:=t_o$;

$h_o:=\Delta t$;

for l=1,2,3 and i=1(1)N:

$y_{li}:=x_{li}^{o}$;

$u_{li} := v^{o}_{li}$;

kon:=<u>false</u>;

$\nu := 0$;

2° $s := 0$;

3° $\beta_s := 2 \times \begin{cases} 1, & \text{if } s=0, \\ 2^{(s+1)/2}, & \text{for } s \text{ odd}, \\ 3 \times 2^{(s-2)/2}, & \text{for } s \text{ even}; \end{cases}$

$\varkappa := 0$;

for $l=1,2,3$ and $i=1(1)N$:

$$z_{li} := y_{li} + \frac{h_o}{\beta_s} u_{li};$$

$$w_{li} := u_{li} - \frac{h_o G}{\beta_s} \sum_{\substack{j=1 \\ j \neq i}}^{N} m_j \frac{y_{li} - y_{lj}}{(\bar{r}_{ij})^3},$$

where $\bar{r}_{ij} = [\sum_{l=1}^{3} (y_{li} - y_{lj})^2]^{1/2}$;

$\varkappa := \max(\varkappa, |z_{li}|, |y_{li}|, |w_{li}|, |u_{li}|)$;

4° for $p=2(1)\beta_s - 1$:

for $l=1,2,3$ and $i=1(1)N$:

$a_{li} := z_{li}$;

$b_{li} := w_{li}$;

for $l=1,2,3$ and $i=1(1)N$:

$$z_{li} := y_{li} + \frac{2h_o}{\beta_s} b_{li};$$

$$w_{li} := u_{li} - \frac{2h_o G}{\beta_s} \sum_{\substack{j=1 \\ j \neq i}}^{N} m_j \frac{a_{li} - a_{lj}}{(\tilde{r}_{ij})^3},$$

where $\tilde{r}_{ij} = [\sum_{l=1}^{3} (a_{li} - a_{lj})^2]^{1/2}$;

$\varkappa := \max(\varkappa, |z_{li}|, |w_{li}|)$;

for $l=1,2,3$ and $i=1(1)N$:

$y_{li} := a_{li}$;

$u_{li} := b_{li}$;

5° for $l=1,2,3$ and $i=1(1)N$:

$p := (l-1)N + i$

$$T_{s0p} := \frac{1}{2}(z_{li} + y_{li}) + \frac{h_o}{\beta_s}[w_{li} + \frac{u_{li}}{2} - \frac{h_o G}{\beta_s} \sum_{\substack{j=1 \\ j \neq i}}^{N} m_j \frac{z_{li} - z_{lj}}{(\bar{r}_{ij})^3}];$$

$$T_{s0,p+3N} := \frac{1}{2}(w_{1i}+u_{1i}) - \frac{h_oG}{\beta_s} \sum_{\substack{j=1\\j\neq i}}^{N} m_j\Big[\frac{z_{1i}-z_{1j}}{(\bar{r}_{ij})^3} + \frac{y_{1i}-y_{1j}+\frac{2h_o}{\beta_s}(w_{1i}-w_{1j})}{2(\tilde{r}_{ij})^3}\Big],$$

$$\text{where } \bar{r}_{ij}=\Big[\sum_{1=1}^{3}(z_{1i}-z_{1j})^2\Big]^{1/2},$$

$$\tilde{r}_{ij}=\Big\{\sum_{1=1}^{3}\Big[y_{1i}-y_{1j}+\frac{2h_o}{\beta_s}(w_{1i}-w_{1j})\Big]^2\Big\}^{1/2};$$

```
6° if s=0 then:
      s:=s+1;
      for l=1,2,3 and i=1(1)N:
         y_li:=x^ν_li;
         u_li:=v^ν_li;
      go to 3°;
   if s>6 then q:=6;
   if s≤6 then q:=s;
7° for k=1(1)q:
      for p=1(1)6N:
         T_s-k,k,p := T_s-k+1,k-1,p + (T_s-k+1,k-1,p - T_s-k,k-1,p)/((β_s/β_s-k)^2 - 1);
      if s<7 then:
         rt:=false;
         for p=1(1)6N:
            if |T_s-k,k,p - T_s-k,k-1,p|/ϰ ≥ ε then rt:=true;
                     ( ε denotes an accuracy given beforehand )
8° if s<7 then:
      if rt=true then go to 9°;
      if rt=false then:
         if kon=true or s>4 then:
            t:=t+h_o;
            go to 10°;
         h_o:=1.5h_o;
         for l=1,2,3 and i=1(1)N:
            y_li:=x^ν_li;
            u_li:=v^ν_li;
         go to 2°;
   for p=1(1)6N:
      if |T_s-6,6,p - T_s-7,6,p|/ϰ ≥ ε then go to 9°;
```

```
    t:=t+h_o;
    h_o:=0.9(0.6)^(s-6) h_o;
    go to 10°;
9°  s:=s+1;
    for l=1,2,3 and i=1(1)N:
       y_li:=x^ν_li;
       u_li:=v^ν_li;
    if s≤9 then go to 3°;
    h:=h_o/2;
    go to 2°;
10° if s>6 then k:=s-6;
     if s≤6 then k:=0;
     for l=1,2,3 and i=1(1)N:
        p:=(l-1)N+i;
        y_li:=x^ν_li:=T_k,s-k,p;
        u_li:=v^ν_li:=T_k,s-k,p+3N;
     if t+h_o>t_o+(ν+1)Δt then:
        h:=h_o;
        kon:=true;
        h_o:=t_o+(ν+1)Δt-t;
        go to 2°;
     if kon=true then:
        printout x^ν_li,v^ν_li (l=1,2,3; i=1(1)N);
        kon:=false;
        h_o:=h;
        ν:=ν+1;
        for l=1,2,3 and i=1(1)N:
           x^ν_li:=y_li;
           v^ν_li:=u_li;
        if ν<n then go to 2°;
        go to 11°;
     go to 2°;
11° stop.
```

Algorithm 8. Rational extrapolation with an automatic step size correction

```
1°,2°,4°-6°,8°-11° as in Algorithm 7;
3°  for p=1(1)6N:
       T_s,-1,p:=0;
    continue as in Algorithm 7;
7°  for k=1(1)q and p=1(1)6N:
```

$A := T_{s-k+1,k-1,p}$;
$B := T_{s-k,k-1,p}$;
$C := T_{s-k+1,k-2,p}$;
$D := (\beta_s/\beta_{s-k})^2(B-C)-A+C$;
if $D=0$ then $T_{s-k,k,p} := 0$;
if $D \neq 0$ then $T_{s-k,k,p} := A+(A-B)(A-C)/D$;
continue as in Algorithm 7.

At the end of the previous section we presented the example of the analytical solution of the two-body problem. Let us now compare the results obtained from the algorithms given above with this solution (see Tables VI-XIII and Figs.6-9). In Algorithms 1 and 2 we have assumed the desired solution accuracy $\varepsilon=10^{-9}$ while in Algorithms 3-8, $\varepsilon=10^{-12}$. Let us pay special attention to the changes of the values of the constants of motion (Fig.9).

Table VI. Solution of the two-body problem by the modified Euler method ($\Delta t=0.01$)

ν	i	l	x_{li}	v_{li}	characteristics constants of motion
20	1	1	0.309019104953	-5.975644728669	$\alpha=200$, $\beta=67$
		2	0.951057627661	1.941625203541	$\Delta=0.000027\%$
	2	1	0.000002075279	0.000017947143	$c_1=-8.5\times10^{-14}$
		2	0.000000917772	0.000013039363	$c_2=6.283185307179$
					$c_4=1.000000000000$
					$c_5=2.7\times10^{-14}$
					$c_9=6.283185307642$
					$c_{10}=-19.73909255090$
60	1	1	-0.809013088761	3.693141931155	$\alpha=194$, $\beta=65$
		2	-0.587770931658	-5.083173315388	$\Delta=0.000181\%$
	2	1	0.000005433157	-0.000011091915	$c_1=-4.72\times10^{-13}$
		2	0.000013087783	0.000034137515	$c_2=6.283185307178$
					$c_4=1.000000000000$
					$c_5=4.43\times10^{-13}$
					$c_9=6.283185308461$
					$c_{10}=-19.73909254575$
100	1	1	1.000000000683	0.000016324808	$\alpha=206$, $\beta=69$
		2	0.000016272576	6.283185305001	$\Delta=0.000260\%$
	2	1	0.000000000000	-0.000000000049	$c_1=1.9\times10^{-14}$
		2	0.000018870756	0.000000000000	$c_2=6.283185307178$
					$c_4=1.000000000000$
					$c_5=-2.03$ 10
					$c_9=6.283185309337$
					$c_{10}=-19.73909254026$
					CPU time 9.48 sec

Table VII. Solution of the two-body problem by the fourth order method of Runge-Kutta ($\Delta t=0.01$)

ν	i	l	x_{li}	v_{li}	characteristics constants of motion
20	1	1	0.309019021021	-5.975646876894	$\alpha=16$, $\beta=3$
		2	0.951057397727	1.941623555593	$\Delta=0.000001\%$
	2	1	0.000002075280	0.000017947149	$c_1=-0.6\times10^{-14}$
		2	0.000000917772	0.000013039368	$c_2=6.283185307180$
					$c_4=1.000000000000$
					$c_5=0.1\times10^{-14}$
					$c_9=6.283185304486$
					$c_{10}=-19.73909257074$
60	1	1	-0.809011102531	3.693155887809	$\alpha=16$, $\beta=3$
		2	-0.587772573025	-5.083168151236	$\Delta=0.000058\%$
	2	1	0.000005433151	-0.000011091957	$c_1=-2.1\times10^{-14}$
		2	0.000013087788	0.000034137499	$c_2=6.283185307180$
					$c_4=1.000000000000$
					$c_5=2.0\times10^{-14}$
					$c_9=6.283185299499$
					$c_{10}=-19.73909260207$
100	1	1	0.999999995914	-0.000005467238	$\alpha=16$, $\beta=3$
		2	0.000019740612	6.283185319938	$\Delta=0.000087\%$
	2	1	0.000000000000	0.000000000016	$c_1=0.6\times10^{-14}$
		2	0.000018870745	0.000000000000	$c_2=6.283185307180$
					$c_4=1.000000000000$
					$c_5=-1.1\times10^{-14}$
					$c_9=6.283185294272$
					$c_{10}=-19.73909263492$
					CPU time 1.33 sec

Notations: ν - moment, i - number of body, l - axis, α - number of calculations of right-hand sides of (2.1.1)-(2.1.2) from the moment ν-1 to the moment ν, β - number of integration steps, Δ - mean relative error, i.e.

$$\Delta = \frac{1}{2}\left(\frac{\|x-x^d\|}{\|x^d\|} + \frac{\|v-v^d\|}{\|v^d\|}\right)\times 100\%,$$

where

$$\|x\| = \sum_{l=1}^{2}\sum_{i=1}^{2} |x_{li}|.$$

From the presented results it follows that Euler's method is not efficient because this method requires a lot of evaluations of the values of function and a large number of steps to get the solution with desired accuracy. The fourth-order Runge-Kutta method has the advantage of simplicity and is efficient for low accuracy calculations although the multistep methods with a constant step size are more efficient taking into account the time of calculations and a greater number of bodies (see Fig.10). Extrapolation methods, especially the method of Gragg-Bulirsch-Stoer, have always done well in comparison with other methods, especially for high accuracy calculations. In Table XIV we present the solution of the six-body problem (the motion of all big planets and the earth) obtained by the rational extrapolation method with accuracy $\varepsilon=10^{-10}$. Let us note that the constants of motion are changed insignificantly.

Table VIII. Solution of the two-body problem by the Adams -Bashforth method of fourth order ($\Delta t=0.01$)

ν	i	l	x_{li}	v_{li}	characteristics constants of motion
20	1	1	0.309024084212	-5.975635969831	$\alpha=1$, $\beta=1$
		2	0.951054843239	1.941651659103	$\Delta=0.000541\%$
	2	1	0.000002075264	0.000017947116	$c_1=-0.3\times10^{-14}$
		2	0.000000917780	0.000013039283	$c_2=6.283185307180$
					$c_4=1.000000000000$
					$c_5=0$
					$c_9=6.283178181868$
					$c_{10}=-19.73913732349$
60	1	1	-0.809007420783	3.693161879678	$\alpha=1$, $\beta=1$
		2	-0.587763609283	-5.083198377822	$\Delta=0.000672\%$
	2	1	0.000005433140	-0.000011091975	$c_1=-0.8\times10^{-14}$
		2	0.000013087761	0.000034137590	$c_2=6.283185307180$
					$c_4=1.000000000000$
					$c_5=0.9\times10^{-14}$
					$c_9=6.283161455409$
					$c_{10}=-19.73924241959$
100	1	1	0.999986789292	-0.000168600535	$\alpha=1$, $\beta=1$
		2	0.000045676303	6.283227731910	$\Delta=0.002394\%$
	2	1	0.000000000040	0.000000000506	$c_1=0.1\times10^{-14}$
		2	0.000018870667	-0.000000000127	$c_2=6.283185307180$
					$c_4=1.000000000000$
					$c_5=-0.2\times10^{-14}$
					$c_9=6.283144312558$
					$c_{10}=-19.73935013317$
					CPU time 0.62 sec

Table IX. Solution of the two-body problem by the Adams -Bashforth method of seventh order ($\Delta t=0.01$)

ν	i	l	x_{li}	v_{li}	characteristics constants of motion
20	1	1	0.309019022204	-5.975646802934	$\alpha=1$, $\beta=1$
		2	0.951057405382	1.941623596724	$\Delta=0.000009\%$
	2	1	0.000002075280	0.000017947149	$c_1=0$
		2	0.000000917772	0.000013039367	$c_2=6.283185307180$
					$c_4=1.000000000000$
					$c_5=0$
					$c_9=6.283185294897$
					$c_{10}=-19.73909263099$
60	1	1	-0.809011809105	3.693160950825	$\alpha=1$, $\beta=1$
		2	-0.587772818427	-5.083164780476	$\Delta=0.000109\%$
	2	1	0.000005433153	-0.000011091972	$c_1=-0.6\times10^{-14}$
		2	0.000013087789	0.000034137489	$c_2=6.283185307180$
					$c_4=1.000000000000$
					$c_5=0.5\times10^{-14}$
					$c_9=6.283190046437$
					$c_{10}=-19.73906277617$
100	1	1	1.006936177450	-0.062625382711	$\alpha=1$, $\beta=1$
		2	-0.001508587904	6.286525069150	$\Delta=0.948314\%$ (?!)
	2	1	-0.000000020832	0.000000188088	$c_1=0$
		2	0.000018875335	-0.000000010031	$c_2=6.283185307180$
					$c_4=1.000000000000$
					$c_5=-0.1\times10^{-14}$
					$c_9=6.330033864676$
					$c_{10}=-19.44415347429$
					CPU time 0.73 sec

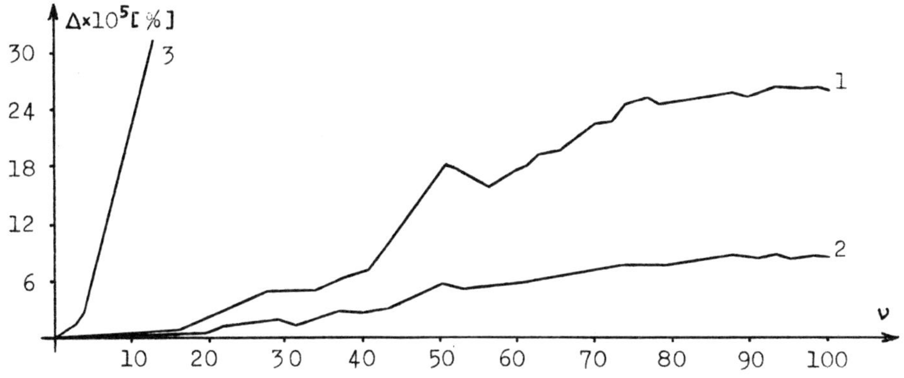

Fig.6. Mean relative errors in the modified Euler method (1), Runge-Kutta method (2), and fourth order method of Adams -Bashforth with constant step size (3)

Table X. Solution of the two-body problem by the predictor -corrector method of seventh order ($\Delta t=0.01$)

ν	i	l	x_{li}	v_{li}	characteristics constants of motion
20	1	1	0.309019022968	-5.975646803356	$\alpha=4$, $\beta=1$
		2	0.951057406588	1.941623608714	$\Delta=0.000008\%$
	2	1	0.000002075280	0.000017947149	$c_1=0.2\times10^{-14}$
		2	0.000000917772	0.000013039367	$c_2=6.283185307180$
					$c_4=1.000000000000$
					$c_5=0$
					$c_9=6.283185307729$
					$c_{10}=-19.73909255036$
60	1	1	-0.809011175289	3.693155381743	$\alpha=4$, $\beta=1$
		2	-0.587772515932	-5.083168334221	$\Delta=0.000049\%$
	2	1	0.000005433151	-0.000011091956	$c_1=-0.7\times10^{-14}$
		2	0.000013087788	0.000034137500	$c_2=6.283185307180$
					$c_4=1.000000000000$
					$c_5=0.6\times10^{-14}$
					$c_9=6.283185309068$
					$c_{10}=-19.73909254195$
100	1	1	1.000000001081	-0.000004610732	$\alpha=4$, $\beta=1$
		2	0.000019604561	6.283185303648	$\Delta=0.000073\%$
	2	1	0.000000000000	0.000000000014	$c_1=0.1\times10^{-14}$
		2	0.000018870746	0.000000000000	$c_2=6.283185307180$
					$c_4=1.000000000000$
					$c_5=0$
					$c_9=6.283185310442$
					$c_{10}=-19.73909253331$
					CPU time 0.83 sec

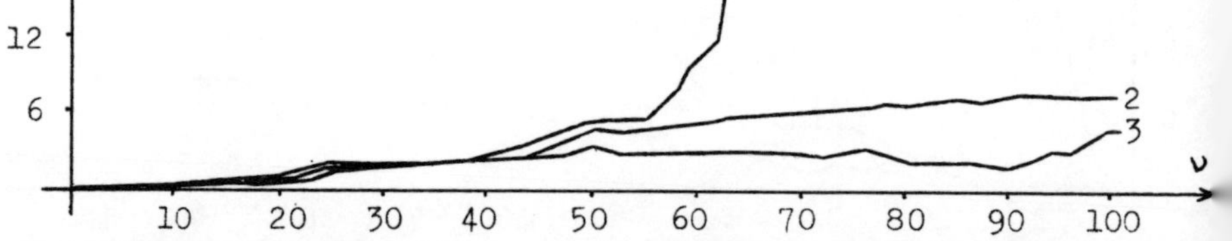

Fig.7. Mean relative errors in the Adams-Bashforth method of seventh order (1), seventh order predictor-corrector method (2) and sixth order method of Butcher (3)

Table XI. Solution of the two-body problem by the Butcher method of sixth order ($\Delta t=0.01$)

ν	i	l	x_{li}	v_{li}	characteristics constants of motion
20	1	1	0.309018995184	-5.975646972454	$\alpha=4$, $\beta=1$
		2	0.951057407897	1.941623674687	$\Delta=0.000010\%$
	2	1	0.000002075280	0.000017947149	$c_1=0$
		2	0.000000917772	0.000013039367	$c_2=6.283185307180$
					$c_4=1.000000000000$
					$c_5=0$
					$c_9=6.283185442777$
					$c_{10}=-19.73909170188$
60	1	1	-0.809011388446	3.693153798670	$\alpha=4$, $\beta=1$
		2	-0.587772483423	-5.083168812277	$\Delta=0.000029\%$
	2	1	0.000005433152	-0.000011091951	$c_1=-0.1\times10^{-14}$
		2	0.000013087788	0.000034137501	$c_2=6.283185307180$
					$c_4=1.000000000000$
					$c_5=-0.5\times10^{-14}$
					$c_9=6.283185728773$
					$c_{10}=-19.73908990486$
100	1	1	1.000000000000	0.000001685289	$\alpha=4$, $\beta=1$
		2	0.000018602899	6.283184578652	$\Delta=0.000044\%$
	2	1	-0.000000000001	-0.000000000005	$c_1=0.2\times10^{-14}$
		2	0.000018870749	0.000000000002	$c_2=6.283185307180$
					$c_4=1.000000000000$
					$c_5=-1.6\times10^{-14}$
					$c_9=6.283186014768$
					$c_{10}=-19.73908810790$
					CPU time 0.63 sec

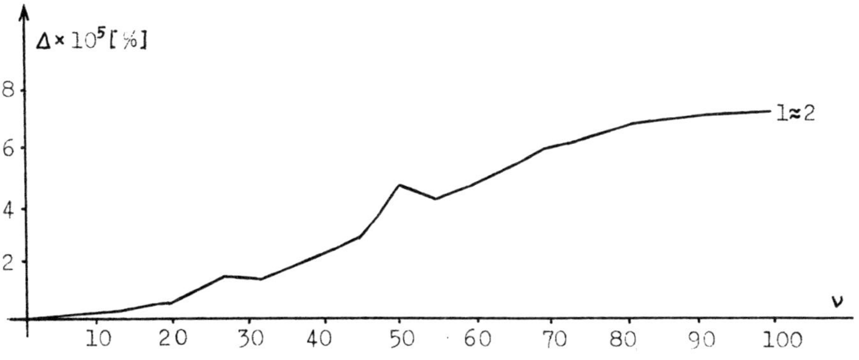

Fig.8. Mean relative errors in polynomial (1) and rational (2) extrapolation

Table XII. Solution of the two-body problem by the polynomial extrapolation method ($\Delta t=0.01$)

ν	i	l	x_{li}	v_{li}	characteristics constants of motion
20	1	1	0.309019022949	-5.975646803227	$\alpha=737$, $\beta=30$
		2	0.951057406548	1.941623608339	$\Delta=0.000009\%$
	2	1	0.000002075280	0.000017947149	$c_1=-1.77\times10^{-13}$
		2	0.000000917772	0.000013039367	$c_2=6.283185307179$
					$c_4=1.000000000000$
					$c_5=1.4\times10^{-14}$
					$c_9=6.283185307177$
					$c_{10}=-19.73909255383$
60	1	1	-0.809011174176	3.693155389497	$\alpha=626$, $\beta=31$
		2	-0.587772516367	-5.083168331252	$\Delta=0.000049\%$
	2	1	0.000005433151	-0.000011091956	$c_1=-9.07\times10^{-13}$
		2	0.000013087788	0.000034137500	$c_2=6.283185307177$
					$c_4=1.000000000000$
					$c_5=7.29\times10^{-13}$
					$c_9=6.283185307172$
					$c_{10}=-19.73909255386$
100	1	1	0.999999999996	-0.000004639994	$\alpha=626$, $\beta=31$
		2	0.000019609225	6.283185307193	$\Delta=0.000074\%$
	2	1	0.000000000000	0.000000000014	$c_1=5.8\times10^{-14}$
		2	0.000018870746	0.000000000000	$c_2=6.283185307176$
					$c_4=0.999999999999$
					$c_5=-6.30\times10^{-13}$
					$c_9=6.283185307169$
					$c_{10}=-19.73909255388$
					CPU time 33.29 sec

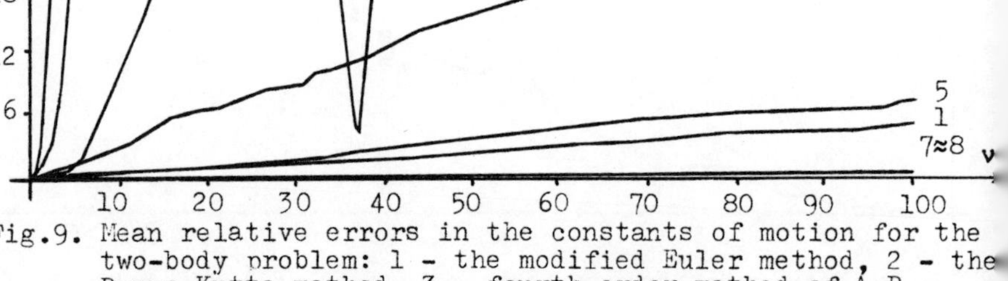

Fig.9. Mean relative errors in the constants of motion for the two-body problem: 1 - the modified Euler method, 2 - the Runge-Kutta method, 3 - fourth order method of A-B, 4 - 7th order method of A-B, 5 - predictor-corrector method, 6 - 6th order method of Butcher, 7 - polynomial extrapolation, 8 - rational extrapolation

Table XIII. Solution of the two-body problem by the rational extrapolation ($\Delta t=0.01$)

ν	i	l	x_{li}	v_{li}	characteristics constants of motion
20	1	1	0.309019022940	-5.975646803275	$\alpha=510$, $\beta=29$
		2	0.951057406546	1.941623608309	$\Delta=0.000009\%$
	2	1	0.000002075280	0.000017947149	$c_1=-1.28\times10^{-13}$
		2	0.000000917772	0.000013039367	$c_2=6.283185307174$
					$c_4=1.000000000000$
					$c_5=5.20\times10^{-13}$
					$c_9=6.283185307185$
					$c_{10}=-19.73909255378$
60	1	1	-0.809011174141	3.693155389776	$\alpha=452$, $\beta=31$
		2	-0.587772516395	-5.083168331150	$\Delta=0.000049\%$
	2	1	0.000005433151	-0.000011091956	$c_1=-6.78\times10^{-13}$
		2	0.000013087788	0.000034137500	$c_2=6.283185307170$
					$c_4=1.000000000006$
					$c_5=-0.000000000002$
					$c_9=6.283185307178$
					$c_{10}=-19.73909255382$
100	1	1	0.999999999997	-0.000004640473	$\alpha=492$, $\beta=27$
		2	0.000019609313	6.283185307208	$\Delta=0.000074\%$
	2	1	0.000000000000	0.000000000014	$c_1=-0.000000000002$
		2	0.000018870746	0.000000000000	$c_2=6.283185307173$
					$c_4=1.000000000005$
					$c_5=-0.000000000002$
					$c_9=6.283185307195$
					$c_{10}=-19.73909255372$
					CPU time 24.54 sec

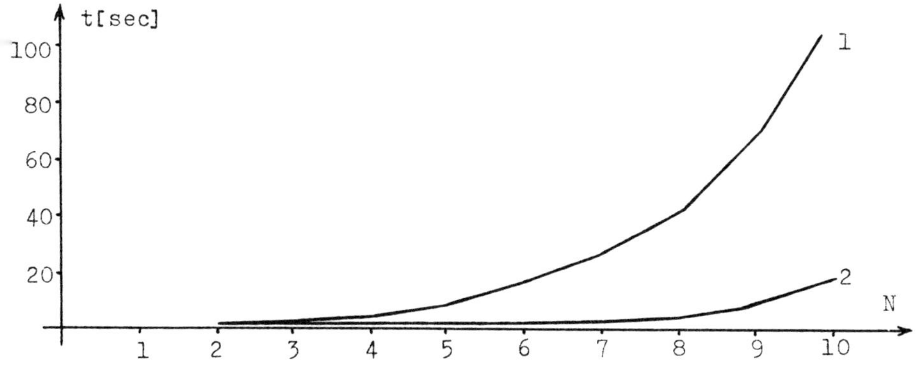

Fig.10. Increase of computatio time according to the number of bodies: 1 - the Runge-Kutta method of fourth order, 2 - fourth order method of Adams-Bashforth

Table XIV. Solution of the six-body problem by the rational extrapolation method ($\Delta t=0.01$, $\varepsilon=0.0000000001$)

$m_1=1.0$ $m_4=14.54$

$m_2=317.84$ $m_5=17.25$

$m_3=95.17$ $m_6=332958.0$

ν	i	l	x_{li}	v_{li}	characteristics constants of motion
0	1	1	-0.259497000000	5.972696300000	c_1=-954.5544775720
		2	-0.980998800000	-1.629881500000	c_2= 117.7002482030
		3	0.000000000000	0.000000000000	c_3= -10.4297036220
	2	1	1.414558600000	-2.680912700000	c_4=-541.5746728230
		2	4.866311400000	0.898923300000	c_5=1451.693675512
		3	-0.051585100000	-0.056515200000	c_6= 14.8874359230
	3	1	-6.802932100000	-1.477600600000	c_7= 19.8337304460
		2	6.139742800000	-1.517166000000	c_8= 111.9337106160
		3	0.165267500000	0.085127000000	c_9=7394.867175399
	4	1	-14.184761000000	0.913893700000	c_{10}=-1447.506534332
		2	-11.994141000000	-1.163498100000	
		3	0.139608300000	-0.016161100000	
	5	1	-7.955768600000	1.096194400000	
		2	-29.214971200000	-0.294359500000	
		3	0.784048200000	-0.019331400000	
	6	1	0.000000000000	0.000000000000	
		2	0.000000000000	0.000000000000	
		3	0.000000000000	0.000000000000	
40	1	1	0.797541107523	-3.846015062066	α=251, β=13
		2	0.591563274813	5.025323936151	c_1=-954.5544775706
		3	-0.000000976194	-0.000006751932	c_2= 117.7002482296
	2	1	0.316844297262	-2.785262940782	c_3= -10.4297036217
		2	5.106515144444	0.300169990909	c_4=-541.5746728185
		3	-0.072775915477	-0.049051702077	c_5=1451.693675505
	3	1	-7.365338029930	-1.332779550352	c_6= 14.8874359230
		2	5.508618667294	-1.636083804869	c_7= 19.8337304487
		3	0.198595669750	0.081406895179	c_8= 111.9337106160
	4	1	-13.812278783563	0.948344507109	c_9=7394.867175423
		2	-12.453555964992	-1.133405423328	c_{10}=-1447.506534270
		3	0.133076196103	-0.016496547734	
	5	1	-7.516402712151	1.100593438472	
		2	-29.329386477818	-0.277705512848	
		3	0.776226732621	-0.019775209985	
	6	1	0.000019665856	0.000085974731	
		2	0.000113764559	0.000583392275	
		3	-0.000001136890	-0.000006023630	
70	1	1	-0.834716613115	-3.464366676587	α=251, β=13
		2	0.528884624490	-5.331126688862	c_1=-954.5544775704
		3	-0.000004456721	-0.000014717422	c_2= 117.7002482271

Table XIV. (cont.)

ν	i	l	x_{li}	v_{li}	characteristics constants of motion
	2	1	-0.519257080209	5.129109236037	c_3 = -10.4297036213
		2	5.129109236037	-0.148168783933	c_4 =-541.5746728187
		3	-0.086471074620	-0.042076110339	c_5 =1451.693675507
	3	1	-7.748088899089	-1.218092491603	c_6 = 14.8874359228
		2	5.005636519252	-1.715739678386	c_7 = 19.8337304489
		3	0.222546466195	0.078208937528	c_8 = 111.9337106160
	4	1	-13.523988098594	0.973493704642	c_9 =7394.867175428
		2	-12.790106582335	-1.110172262470	c_{10}=-1447.506534227
		3	0.128090954515	-0.016736761872	
	5	1	-7.185750880984	1.103728611823	
		2	-29.410818907312	-0.265171261328	
		3	0.770244611979	-0.020105176924	
	6	1	0.000042321224	0.000042511391	
		2	0.000361118319	0.001063582780	
		3	-0.000003779140	-0.000011740807	
100	1	1	-0.259016356162	5.973304163620	α=174, β=12
		2	-0.980382118775	-1.625497960646	c_1 =-954.5544775694
		3	-0.000008452967	-0.000012215906	c_2 = 117.7002462307
	2	1	-1.342026032227	-2.697120365948	c_3 = -10.4297036213
		2	5.019013993211	-0.582554112054	c_4 =-541.5746728194
		3	-0.097929676839	-0.034181641068	c_5 =1451.693675503
	3	1	-8.095746401645	-1.098934310872	c_6 = 14.8874359227
		2	4.480021356115	-1.786932576785	c_7 = 19.8337304491
		3	0.245488434425	0.074685035122	c_8 = 111.9337106159
	4	1	-13.228242976545	0.998038763826	c_9 =7394.867175425
		2	-13.119604311790	-1.086389001347	c_{10}=-1447.506534197
		3	0.123035124032	-0.016967125784	
	5	1	-6.854179662278	1.106722613446	
		2	-29.488485660718	-0.252601270895	
		3	0.764163879045	-0.020432618106	
	6	1	0.000035214223	-0.000096984950	
		2	0.000745447444	0.001485774742	
		3	-0.000008259849	-0.000018242567	

CPU time 109.89 sec

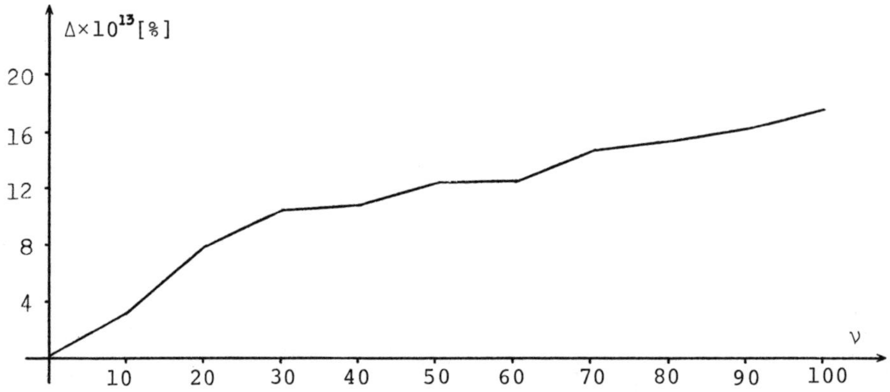

Fig.11. Mean relative errors in the constants of motion for the six-body problem

2.3. DISCRETE MECHANICS OF GREENSPAN

From the theoretical justifications and examples mentioned in the previous section, it follows that conventional numerical methods applied to the N-body problem do not conserve the constants of motion. For exapmle, it can be shown that the calculated total energy has an error of the same order as velocities of material points. Thus, a problem arises to construct such numerical methods that conserve the constants of motion exactly. In other words, we would like to obtain numerical methods that give the solutions with the same properties as follows from the analytical theory. We number, apart from the conservation of constants of motion, the invariability of discrete equations with respect to some transformations of the frame of reference among these properties. A numerical method that has such properties we will call a discrete mechanics.

The purpose of this section is to present Greenspan's discrete mechanics ([36],[37],[38],[39],[40],[42]). Discrete dynamical equations of this mechanics fulfil the properties mentioned above.

2.3.1. BASIC RELATIONS AND THEIR PROPERTIES

Let F^k_{li} ($l=1,2,3$; $i=1(1)N$; $k=0,1,2,\ldots$) denote the l-th component of the gravitational force acting on material point P_i of mass m_i at the moment t_k, r^k_{ij} is the distance between P_i and P_j ($i\neq j$) and x^k_{li} denotes the l-th coordinate of P_i at t_k in an inertial frame of reference. In [38] (see also [37],[42],[65]) D.Greenspan has defined F^k_{li} by the formula

$$F^k_{li} = -Gm_i \sum_{\substack{j=1\\ j\neq i}}^{N} m_j \frac{x^{k+1}_{li}+x^k_{li}-x^{k+1}_{lj}-x^k_{lj}}{r^{k+1}_{ij} r^k_{ij}(r^{k+1}_{ij}+r^k_{ij})} ; \qquad (2.3.1)$$

$$l=1,2,3;\ i=1(1)N;\ k=0,1,2,\ldots;$$

where G denotes the gravitational constant and

$$r^k_{ij} = \left[\sum_{l=1}^{3} (x^k_{li}-x^k_{lj})^2 \right]^{1/2}.$$

Let us note that the discrete form (2.3.1) of Newtonian gravitational force can be obtained from (2.1.2) by substitutions $x_i=x^k_i$, $r_{ij}=r^k_{ij}$ and taking into account the following relationships

$$\begin{aligned} r^{k+1} + r^k &= 2r^k,\\ x^{k+1} + x^k &= 2x^k, \qquad (2.3.2)\\ r^k r^{k+1} &= (r^k)^2. \end{aligned}$$

These relationships are obvious if we leave out of account the terms converging to zero as $\Delta t=t_{k+1}-t_k\to 0$. In (2.3.2) the lower indices are omitted as inessential. Moreover, let us note that

$$F^k_{li} \neq [F_{li}(t_k)+F_{li}(t_{k+1})]/2 \qquad (2.3.3)$$

and

$$F^k_{li} \to F_{li}(t_k) \text{ as } \Delta t\to 0,$$

where F_{li} is given by (2.1.2).

For an isolated system of N material points, the dynamical difference equations of Greenspan's discrete mechanics are of the form ([36],[37],[38],[39],[42])

$$\frac{v_{li}^{k+1}+v_{li}^{k}}{2} = \frac{x_{li}^{k+1}-x_{li}^{k}}{\Delta t}, \tag{2.3.4}$$

$$\frac{F_{li}^{k}}{m_i} = \frac{v_{li}^{k+1}-v_{li}^{k}}{\Delta t}; \tag{2.3.5}$$

$$l=1,2,3;\ i=1(1)N;\ k=0,1,2,\ldots;$$

where $\Delta t=t_{k+1}-t_k$ and F_{li}^k is given by (2.3.1). The symmetry of these equations with respect to translation, rotation and under uniform motion of the frame of reference has been shown in [40].

It appears that the equations (2.3.4)-(2.3.5) conserve the constants of motion. Namely, we have the following theorems:

Theorem 2.1. If the motion of an isolated system of N material points in an inertial frame of reference is described by (2.3.1), (2.3.4)-(2.3.5), then for each k=0,1,2,...

$$\sum_{i=1}^{N} m_i v_{li}^{k} = c_l, \tag{2.3.6}$$

$$\sum_{i=1}^{N} m_i x_{li}^{k} = c_l t_k + c_{l+3}; \quad l=1,2,3; \tag{2.3.7}$$

where c_l and c_{l+3} are constants.

Proof. To prove (2.3.6) it is sufficient to show that for each k=0,1,2,...

$$\sum_{i=1}^{N} m_i (v_i^{k+1}-v_i^{k}) = 0, \tag{2.3.8}$$

where $v_i^s=[v_{1i}^s, v_{2i}^s, v_{3i}^s]$ (s=k or s=k+1). From (2.3.5) and (2.3.1) we have

$$\sum_{i=1}^{N} m_i (v_i^{k+1}-v_i^{k}) = \Delta t \sum_{i=1}^{N} m_i \frac{v_i^{k+1}-v_i^{k}}{\Delta t} = \Delta t \sum_{i=1}^{N} F_i^k$$

$$= -G\Delta t \sum_{i=1}^{N} \sum_{\substack{j=1 \\ j\neq i}}^{N} m_i m_j \frac{x_i^{k+1}+x_i^{k}-x_j^{k+1}-x_j^{k}}{r_{ij}^{k+1} r_{ij}^{k} (r_{ij}^{k+1}+r_{ij}^{k})},$$

where $x_i^s=[x_{1i}^s, x_{2i}^s, x_{3i}^s]$. Since for each component of the form $\dfrac{x_i^{k+1}+x_i^{k}-x_j^{k+1}-x_j^{k}}{r_{ij}^{k+1} r_{ij}^{k} (r_{ij}^{k+1}+r_{ij}^{k})}$ we can find the component $\dfrac{x_j^{k+1}+x_j^{k}-x_i^{k+1}-x_i^{k}}{r_{ji}^{k+1} r_{ji}^{k} (r_{ji}^{k+1}+r_{ji}^{k})}$, and $r_{ij}^s = r_{ji}^s$ (s=k or s=k+1), then the relation (2.3.8) is proved. To prove (2.3.7) it is sufficient to show that for each k=0,1,2,... we have

$$\sum_{i=1}^{N} m_i (x_i^{k+1} - x_i^k) = C\Delta t,$$

where $x_i^s = [x_{1i}^s, x_{2i}^s, x_{3i}^s]$ (s=k or s=k+1), $C=[c_1,c_2,c_3]$, $\Delta t = t_{k+1} - t_k$. But (2.3.4) yields

$$\sum_{i=1}^{N} m_i (x_i^{k+1} - x_i^k) = \frac{\Delta t}{2} \sum_{i=1}^{N} m_i (v_i^{k+1} + v_i^k)$$

$$= \Delta t \left(\frac{1}{2} \sum_{i=1}^{N} m_i v_i^{k+1} + \frac{1}{2} \sum_{i=1}^{N} m_i v_i^k \right).$$

Hence, taking into account (2.3.6), we get (2.3.9).

Theorem 2.2. For the discrete mechanics formulas (2.3.1), (2.3.4)-(2.3.5) the law of conservation of angular momentum is fulfiled, i.e. for each k=0,1,2,... we have

$$\sum_{i=1}^{N} m_i (x_i^k \times v_i^k) = \bar{C} \tag{2.3.10}$$

or, in the scalar form,

$$\sum_{i=1}^{N} m_i (x_{pi}^k v_{qi}^k - x_{qi}^k v_{pi}^k) = c_{l+6},$$

where l=1 for (p,q)=(2,3), l=2 for (p,q)=(3,1), l=3 for (p,q)=(1,2), c_{l+6}=const, $\bar{C}=[c_7,c_8,c_9]$, a×b denotes a vector product of a and b.

Proof. The equation (2.3.10) will be fulfiled if for each k=0,1,2,...

$$\sum_{i=1}^{N} m_i (x_i^{k+1} \times v_i^{k+1} - x_i^k \times v_i^k) = 0. \tag{2.3.11}$$

Taking into account (2.3.4) and (2.3.5), we obtain

$$\sum_{i=1}^{N} m_i (x_i^{k+1} \times v_i^{k+1} - x_i^k \times v_i^k) = \sum_{i=1}^{N} m_i \left[\Delta t x_i^{k+1} \times \left(\frac{v_i^{k+1} - v_i^k}{\Delta t}\right) + (x_i^{k+1} - x_i^k) \times v_i^k\right]$$

$$= \sum_{i=1}^{N} m_i \left[\frac{\Delta t}{m_i} x_i^{k+1} \times F_i^k + (x_i^{k+1} - x_i^k) \times v_i^k\right]. \tag{2.3.12}$$

Since $v_i^k = v_i^{k+1} - \Delta t F_i^k / m_i$ and $v_i^k = 2(x_i^{k+1} - x_i^k)/\Delta t - v_i^{k+1}$, then we have

$$v_i^k = \frac{1}{\Delta t}(x_i^{k+1}-x_i^k) - \frac{\Delta t}{2m_i} F_i^k. \qquad (2.3.13)$$

Substituting (2.3.13) into (2.3.12), we get

$$\sum_{i=1}^{N} m_i(x_i^{k+1}\times v_i^{k+1}-x_i^k\times v_i^k) = \sum_{i=1}^{N} m_i[\frac{\Delta t}{m_i} x_i^{k+1}\times F_i^k - \frac{\Delta t}{2m_i}(x_i^{k+1}-x_i^k)\times F_i^k]$$

$$= \frac{\Delta t}{2} \sum_{i=1}^{N} (x_i^{k+1}+x_i^k)\times F_i^k.$$

Thus, taking into account (2.3.1), we have

$$\sum_{i=1}^{N} m_i(x_i^{k+1}\times v_i^{k+1}-x_i^k\times v_i^k) = -\frac{G\Delta t}{2}\sum_{i=1}^{N}\sum_{\substack{j=1\\ j\neq i}}^{N} \frac{m_i m_j}{r_{ij}^{k+1} r_{ij}^k (r_{ij}^{k+1}+r_{ij}^k)} \times (x_i^{k+1}+x_i^k)\times(x_i^{k+1}+x_i^k-x_j^{k+1}-x_j^k),$$

$$= \frac{G\Delta t}{2}\sum_{i=1}^{N}\sum_{\substack{j=1\\ j\neq i}}^{N} \frac{m_i m_j}{r_{ij}^{k+1} r_{ij}^k (r_{ij}^{k+1}+r_{ij}^k)}(x_i^{k+1}+x_i^k)\times(x_j^{k+1}+x_j^k)=0,$$

because for each component of the form $\frac{(x_i^{k+1}+x_i^k)\times(x_j^{k+1}+x_j^k)}{r_{ij}^{k+1} r_{ij}^k (r_{ij}^{k+1}+r_{ij}^k)}$ we can find the component $\frac{(x_j^{k+1}+x_j^k)\times(x_i^{k+1}+x_i^k)}{r_{ij}^{k+1} r_{ij}^k (r_{ij}^{k+1}+r_{ij}^k)}$.

Theorem 2.3. The discrete mechanics formulas (2.3.1), (2.3.4) -(2.3.5) have an energy conserving property for the isolated system of N material points, i.e. for each k=0,1,2,...

$$\frac{1}{2}\sum_{i=1}^{N} m_i[\sum_{l=1}^{3} (v_{li}^k)^2 - G\sum_{\substack{j=1\\ j\neq i}}^{N} \frac{m_j}{r_{ij}^k}] = c_{10}, \qquad (2.3.14)$$

where c_{10}=const.

Proof. First it will be proved that

$$\sum_{i=1}^{N} F_i^k\circ(x_i^{k+1}-x_i^k) = V^{k+1}-V^k, \qquad (2.3.15)$$

where a∘b denotes a scalar product of some vectors a and b, and

$$V^s = \frac{G}{2}\sum_{i=1}^{N}\sum_{\substack{j=1\\ j\neq i}}^{N} \frac{m_i m_j}{r_{ij}^s}, \quad s=k \text{ or } s=k+1.$$

Of course, the last formula presents a potential energy of the system of N material points at the moment t_s (compare (2.1.13)). Moreover,

$$\sum_{i=1}^{N} F_i^k \circ (x_i^{k+1}-x_i^k) \simeq \int_{t_k}^{t_{k+1}} \sum_{i=1}^{N} F_i \circ dx_i .$$

In other words, the formula (2.3.15) can be interpreted as a discrete approximation of the law that a work done on an isolated system of material points in the interval $[t_k, t_{k+1}]$ is equivalent to the difference between potential energies at t_{k+1} and t_k.

The formula (2.3.15) will be proved by induction. For N=2, using (2.3.1), we have

$$\sum_{i=1}^{2} F_i^k \circ (x_i^{k+1}-x_i^k) = -Gm_1m_2\left[\frac{(x_1^{k+1}+x_1^k-x_2^{k+1}-x_2^k)\circ(x_1^{k+1}-x_1^k)}{r_{12}^{k+1}r_{12}^k(r_{12}^{k+1}+r_{12}^k)} - \frac{(x_2^{k+1}+x_2^k-x_1^{k+1}-x_1^k)\circ(x_2^{k+1}-x_2^k)}{r_{21}^{k+1}r_{21}^k(r_{21}^{k+1}+r_{21}^k)}\right].$$

Since $r_{12}^s = r_{21}^s$ (s=k or s=k+1), then

$$\sum_{i=1}^{2} F_i^k \circ (x_i^{k+1}-x_i^k) = -\frac{Gm_1m_2}{r_{12}^{k+1}r_{12}^k(r_{12}^{k+1}+r_{12}^k)}[(x_1^{k+1})^2+(x_2^{k+1})^2 -2x_1^{k+1}\circ x_2^{k+1}-(x_1^k)^2-(x_2^k)^2+2x_1^k\circ x_2^k]$$

$$= -\frac{Gm_1m_2}{r_{12}^{k+1}r_{12}^k(r_{12}^{k+1}+r_{12}^k)}[(r_{12}^{k+1})^2-(r_{12}^k)^2]$$

$$= -\frac{Gm_1m_2(r_{12}^{k+1}-r_{12}^k)}{r_{12}^{k+1}r_{12}^k}$$

$$= \frac{Gm_1m_2}{r_{12}^{k+1}} - \frac{Gm_1m_2}{r_{12}^k} = V^{k+1}-V^k.$$

Let us now assume that (2.3.15) is true for N=L>2. From (2.3.15), for N=L+1 we have

$$\sum_{i=1}^{L+1} F_i^k \circ (x_i^{k+1}-x_i^k) = -G\sum_{i=1}^{L+1}\sum_{\substack{j=1\\ j\neq i}}^{L+1} \frac{m_im_j}{r_{ij}^{k+1}r_{ij}^k(r_{ij}^{k+1}+r_{ij}^k)} \times (x_i^{k+1}+x_i^k-x_j^{k+1}-x_j^k)\circ(x_i^{k+1}-x_i^k)$$

$$= -G \sum_{i=1}^{L} \sum_{\substack{j=1 \\ j\neq i}}^{L} \frac{m_i m_j}{r_{ij}^{k+1} r_{ij}^{k} (r_{ij}^{k+1} + r_{ij}^{k})} (x_i^{k+1} + x_i^k - x_j^{k+1} - x_j^k) \circ (x_i^{k+1} - x_i^k)$$

$$-G \sum_{i=1}^{L} \frac{m_i m_{L+1}}{r_{i,L+1}^{k+1} r_{i,L+1}^{k} (r_{i,L+1}^{k+1} + r_{i,L+1}^{k})} \times (x_i^{k+1} + x_i^k - x_{L+1}^{k+1} - x_{L+1}^k) \circ (x_i^{k+1} - x_i^k)$$

$$-G \sum_{j=1}^{L} \frac{m_{L+1} m_j}{r_{L+1,j}^{k+1} r_{L+1,j}^{k} (r_{L+1,j}^{k+1} + r_{L+1,j}^{k})} \times (x_{L+1}^{k+1} + x_{L+1}^k - x_j^{k+1} - x_j^k) \circ (x_{L+1}^{k+1} - x_{L+1}^k).$$

Denoting the second sum by A, the third one by B and taking into account our inductive assumption, we can write

$$\sum_{i=1}^{L+1} F_i^k \circ (x_i^{k+1} - x_i^k) = \frac{G}{2} \sum_{i=1}^{L} \sum_{\substack{j=1 \\ j\neq i}}^{L} m_i m_j \left(\frac{1}{r_{ij}^{k+1}} - \frac{1}{r_{ij}^{k}}\right) + A + B. \tag{2.3.16}$$

Now, let us change the index j in B into i and consider the sum A+B. Since $r_{ij}^s = r_{ji}^s$ (s=k or s=k+1), we have

$$A+B = -G \sum_{i=1}^{L} \frac{m_i m_{L+1}}{r_{i,L+1}^{k+1} r_{i,L+1}^{k} (r_{i,L+1}^{k+1} + r_{i,L+1}^{k})} [(x_i^{k+1})^2 + (x_{L+1}^{k+1})^2 - 2x_i^{k+1} \circ x_{L+1}^{k+1} - (x_i^k)^2 - (x_{L+1}^k)^2]$$

$$= G \sum_{i=1}^{L} m_i m_{L+1} \left(\frac{1}{r_{i,L+1}^{k+1}} - \frac{1}{r_{i,L+1}^{k}}\right).$$

Substitution of the last result in (2.3.16) yields

$$\sum_{i=1}^{L+1} F_i^k \circ (x_i^{k+1} - x_i^k) = \frac{G}{2} \sum_{i=1}^{L} \sum_{\substack{j=1 \\ j\neq i}}^{L} m_i m_j \left(\frac{1}{r_{ij}^{k+1}} - \frac{1}{r_{ij}^{k}}\right)$$

$$+ G \sum_{i=1}^{L} m_i m_{L+1} \left(\frac{1}{r_{i,L+1}^{k+1}} - \frac{1}{r_{i,1+1}^{k}}\right)$$

$$= \frac{G}{2} \sum_{i=1}^{L+1} \sum_{\substack{j=1 \\ j\neq i}}^{L+1} m_i m_j \left(\frac{1}{r_{ij}^{k+1}} - \frac{1}{r_{ij}^{k}}\right).$$

This means that (2.3.15) is true for N=L+1. Using mathematical induction, we see that identity (2.3.15) holds for each

N=2,3,4,... .

Now, let us prove that

$$\sum_{i=1}^{N} F_i^k \circ (x_i^{k+1}-x_i^k) = K^{k+1}-K^k, \tag{2.3.17}$$

where

$$K^s = \frac{1}{2}\sum_{i=1}^{N} m_i \sum_{l=1}^{3} (v_{li}^s)^2; \quad s=k \text{ or } s=k+1.$$

The formula (2.3.17) is discrete approximation of the law stating that a work done on an isolated system of material points in the time interval $[t_k,t_{k+1}]$ is equivalent to the difference of kinetic energies at t_{k+1} and t_k. From (2.3.4) and (2.3.5) we have

$$\sum_{i=1}^{N} F_i^k \circ (x_i^{k+1}-x_i^k) = \frac{1}{2}\sum_{i=1}^{N} m_i (v_i^{k+1}-v_i^k)\circ(v_i^{k+1}+v_i^k)$$

$$= \frac{1}{2}\sum_{i=1}^{N} m_i [(v_i^{k+1})^2-(v_i^k)^2]$$

$$= \frac{1}{2}\sum_{i=1}^{N} m_i \sum_{l=1}^{3} (v_{li}^{k+1})^2 - \frac{1}{2}\sum_{i=1}^{N} m_i \sum_{l=1}^{3} (v_{li}^k)^2$$

$$= K^{k+1}-K^k,$$

and so (2.3.17) is proved. Finally, from (2.3.15) and (2.3.17) we obtain that for each k=0,1,2,...

$$K^{k+1}-V^{k+1} = K^k-V^k.$$

Hence

$$K^k-V^k = \text{const.}$$

It is worthwhile to notice that in order to assure the truthfulness of (2.3.10) and (2.3.14), i.e. to prove the constancy of angular momentum and energy, the form of discrete Newtonian gravitational force is fundamental.

2.3.2. STABILITY AND CONVERGENCE

In this section we will be engaged in numerical analysis of the discrete mechanics method. First of all let us note that because of (2.3.3) we can not use the well-known theorems regarding the consistency, stability and convergence of symmetric one-step methods ([86]). Our analysis will be carried out based on the notions from Sect.1.2.

Without loss of generality let us assume that a discrete mechanics method be used to solve the N-body problem in the time interval [0,1] and let $\Delta t=1/n$. Let us denote $y_p \equiv x_{li}$, $y_{p+3N} \equiv v_{li}$, where $l=[\frac{p-1}{N}]+1$, $i=p-[\frac{p-1}{N}]N$ ([.] denotes an entier), $f_p(y) \equiv F_{li}/m_i$, where F_{li} is given by (2.1.2) and let $E = \underset{6N}{\times} C^{(1)}[0,1]$, $E^o = \underset{6N}{\times} R \times \underset{6N}{\times} C^{(1)}[0,1]$, where the norms in E and E^o are determined by (1.1.5). The problem (2.1.1)-(2.1.2) can now be written in the form

$$\Phi y = \begin{cases} y_p(0) - z_p^o, \\ y_{p+3N}(0) - z_{p+3N}^o, \\ y_p' - y_{p+3N}, \\ y_{p+3N}' - f_p(y); \quad p=1(1)3N; \end{cases} \tag{2.3.18}$$

where $z_p^o \equiv x_{li}^o$, $z_{p+3N}^o \equiv v_{li}^o$ and $\Phi y \in E^o$ for $y \in E$. A discrete initial value problem corresponding to (2.3.18) has the form (compare (2.3.4)-(2.3.5))

$$\Phi_n \eta(\tfrac{\nu}{n}) = \begin{cases} \eta_p(0) - z_p^o, \quad \text{for } \nu=0, \\ \eta_{p+3N}(0) - z_{p+3N}^o, \quad \text{for } \nu=0, \\ \dfrac{\eta_p(\frac{\nu}{n}) - \eta_p(\frac{\nu-1}{n})}{1/n} - [\eta_{p+3N}(\tfrac{\nu}{n}) + \eta_{p+3N}(\tfrac{\nu-1}{n})], \\ \qquad \text{for } \nu=1(1)n, \\ \dfrac{\eta_{p+3N}(\frac{\nu}{n}) - \eta_{p+3N}(\frac{\nu-1}{n})}{1/n} - F_p[\eta(\tfrac{\nu}{n}), \eta(\tfrac{\nu-1}{n})], \\ \qquad \text{for } \nu=1(1)n; \end{cases} \tag{2.3.19}$$

where $F_p \equiv F_{li}^{\nu-1}/m_i$ ($F_{li}^{\nu-1}$ is given by (2.3.1)) depends only on η_q (q=1(1)3N) and is independent of η_{q+3N}. Of course, $\Phi_n \eta \in E_n^o$ for $\eta \in E_n$. In the spaces E_n and E_n^o the norms are determined by (1.2.1).

<u>Theorem 2.4.</u> The discrete mechanics method (2.3.19) is consistent with the initial value problem (2.3.18).

Proof. We ought to show that

$$\lim_{n\to\infty} \| \phi_n(\Phi)\Delta_n y-\Delta_n^o\Phi y \|_{E_n^o} = 0,$$

where Δ_n: $E\to E_n$, Δ_n^o: $E^o\to E_n^o$ denote linear operators and $\phi_n(\Phi)=\Phi_n$. Let

$$\begin{aligned}(\Delta_n y)(\tfrac{\nu}{n}) &= y(\tfrac{\nu}{n}) \text{ for } y\epsilon E,\\ (\Delta_n^o d)(\tfrac{\nu}{n}) &= \begin{cases} d^o, \text{ if } \nu=0, \\ d(\frac{\nu-1}{n}), \text{ if } \nu=1(1)n \text{ and for } d\epsilon E^o.\end{cases}\end{aligned} \tag{2.3.20}$$

Let us consider the difference $\phi_n(\Phi)\Delta_n y-\Delta_n^o\Phi y$. From (2.3.18), (2.3.19) and (2.3.20) we have

$$\phi_n(\Phi)\Delta_n y(\tfrac{\nu}{n})-\Delta_n^o\Phi y(\tfrac{\nu}{n})=\begin{cases} 0, \quad \nu=0, \\ y_p'(\tilde{t}^p)-y'(\frac{\nu-1}{n})-[y_{p+3N}(\frac{\nu}{n})-y_{p+3N}(\frac{\nu-1}{n})]/2, \\ \qquad \nu=1(1)n, \\ y_{p+3N}'(\tilde{\tilde{t}}^p)-y_{p+3N}'(\frac{\nu-1}{n})- F_p[y(\frac{\nu}{n}),y(\frac{\nu-1}{n})] \\ \qquad -f_p[y(\frac{\nu-1}{n})] \quad , \quad \nu=1(1)n; \end{cases} \tag{2.3.21}$$

$$p=1(1)3N;$$

where $\tilde{t}^p,\tilde{\tilde{t}}^p\epsilon(\frac{\nu-1}{n},\frac{\nu}{n})$ denote some intermediate points. From (2.3.1) and (2.1.2) it follows that

$$| F_p[y(\tfrac{\nu}{n}),y(\tfrac{\nu-1}{n})]-f_p[y(\tfrac{\nu-1}{n})] | \to 0 \text{ as } n\to\infty.$$

Moreover,

$$| y_p'(\tilde{t}^p)-y_p'(\tfrac{\nu-1}{n}) | \to 0 \text{ as } n\to\infty,$$

$$| y_{p+3N}'(\tilde{\tilde{t}}^p)-y_{p+3N}'(\tfrac{\nu-1}{n}) | \to 0 \text{ as } n\to\infty,$$

$$| y_{p+3N}(\tfrac{\nu}{n})-y_{p+3N}(\tfrac{\nu-1}{n}) | \to 0 \text{ as } n\to\infty.$$

Hence, from (2.3.21) and from the definition of the norm in E_n^o we have

$$\begin{aligned}||\phi_n(\Phi)\Delta_n y-\Delta_n^o\Phi y||_{E_n^o} &= \max_{\nu=1(1)n} \Big(\sum_{p=1}^{6N} |\phi_n(\Phi)\Delta_n y(\tfrac{\nu}{n})-\Delta_n^o\Phi y(\tfrac{\nu}{n})| \Big)\\ &\le \max_{\nu=1(1)n} \Big\{ \sum_{p=1}^{3N} \big(|y_p'(\tilde{t}^p)-y_p'(\tfrac{\nu-1}{n})|+ \tfrac{1}{2}|y_{p+3N}(\tfrac{\nu}{n})-y_{p+3N}(\tfrac{\nu-1}{n})| \\ &\quad +|y_{p+3N}'(\tilde{t}^p)-y_{p+3N}(\tfrac{\nu-1}{n})| \\ &\quad +|F_p[y(\tfrac{\nu}{n}),y(\tfrac{\nu-1}{n})]-f_p[y(\tfrac{\nu-1}{n})]|\big)\Big\} \to 0\end{aligned}$$

as $n\to\infty$.

Before we show that the discrete mechanics method (2.3.19) is of the first order, i.e. it is consistent with (2.3.18) with the first order (see Def.1.3), we will prove two lemmas. We will also use these lemmas later to prove the stability of the method (2.3.19).

So, let us consider a function $F_p[x(t),y(t)]$, where x and y denote 3N-dimensional vectors and $y(t)=x(t+h)$, where h is some fixed real number such that $t+h\in[0,1]$ if $t\in[0,1]$. Let this function be defined by

$$F_p[x(t),y(t)] \equiv F_{li}[x(t),y(t)] = -G \sum_{\substack{j=1\\ j\neq i}}^{N} m_j \frac{y_{li}+x_{li}-y_{lj}-x_{lj}}{r_{ij}R_{ij}(r_{ij}+R_{ij})} \tag{2.3.22}$$

where $r_{ij}=[\sum_{l=1}^{3}(x_{li}-x_{lj})^2]^{1/2}$, $R_{ij}=[\sum_{l=1}^{3}(y_{li}-y_{lj})^2]^{1/2}$.

Let us note that the form of F_p is analogous to the discrete Newtonian gravitational force (2.3.1), but now this is a function of continuous variables.

<u>Lemma 2.1.</u> If there exist the following constants

$$0<r=\min_{\substack{t\in[0,1]\\ i,j=1(1)N\\ i\neq j}} r_{ij}(t),\quad R=\max_{\substack{t\in[0,1]\\ i,j=1(1)N\\ i\neq j}} r_{ij}(t),\quad M=\max_{i=1(1)N} m_i \tag{2.3.23}$$

then for arbitrary p,q=1(1)3N

$$\left.\begin{array}{l} \left|\dfrac{\partial F_p}{\partial y_q}\right| \\[2ex] \left|\dfrac{\partial F_p}{\partial y_q}\right| \end{array}\right\} \leq \frac{2MNG}{r^3}\,. \tag{2.3.24}$$

Proof. The above inequality can be easily obtained after differentiation of (2.3.22), taking into consideration (2.3.23).

<u>Lemma 2.2.</u> If there exist the constants (2.3.23), then there exists a constant $\tilde{M}\geq 1/3N$ such that

$$|F_p(\bar{x},\bar{y})-F_p(x,y)| \leq \tilde{M} \sum_{q=1}^{3N}(|\bar{x}_q-x_q|+|\bar{y}_q-y_q|);\quad p=1(1)3N. \tag{2.3.25}$$

Proof. Applying Taylor's formula to the function F_p given by (2.3.22), we have

$$F_p(\bar{x},\bar{y})=F_p(x,y)+\sum_{q=1}^{3N}\left[\frac{\partial F_p(\xi)}{\partial x_q}(\bar{x}_q-x_q)+\frac{\partial F_p(\eta)}{\partial y_q}(\bar{y}_q-y_q)\right],$$

where ξ and η denote some intermediate points belonging to the intervals $(x,\bar{x})$ and $(y,\bar{y})$ respectively. If there exist the constants (2.3.23), then the partial derivatives are bounded by some constant (see Lemma 2.1). Denoting this constant by $\tilde{M}_1$, we have

$$|F_p(\bar{x},\bar{y})-F_p(x,y)| \le \tilde{M}_1 \sum_{q=1}^{3N} (|\bar{x}_q-x_q|+|\bar{y}_q-y_q|).$$

Hence, defining the constant $\tilde{M}$ as follows

$$\tilde{M} = \max(\tilde{M}_1, 1/3N),$$

we see that the inequality (2.3.25) is fulfilled.

Remark! The condition $\tilde{M}\geq 1/3N$ is essential to prove the stability of the method (2.3.19). We will find it profitable later.

Now we can prove

Theorem 2.5. If there exist the constants (2.3.23), then the method (2.3.19) is of the first order.

Proof. From Taylor's formula we have

$$y_p(\tfrac{\nu}{n}) = y_p(\tfrac{\nu-1}{n})+\frac{1}{n}y_p'(\tfrac{\nu-1}{n})+\frac{1}{n^2}y_p''(\xi^p), \text{ where } \xi^p\epsilon(\tfrac{\nu-1}{n},\tfrac{\nu}{n}),$$

$$y_{p+3N}(\tfrac{\nu}{n}) = y_{p+3N}(\tfrac{\nu-1}{n})+\frac{1}{n}y_{p+3N}'(\tfrac{\nu-1}{n})+\frac{1}{n^2}y_{p+3N}''(\xi_1^p),$$

$$\text{where } \xi_1^p\epsilon(\tfrac{\nu-1}{n},\tfrac{\nu}{n})$$

or

$$y_{p+3N}(\tfrac{\nu}{n}) = y_{p+3N}(\tfrac{\nu-1}{n})+\frac{1}{n}y_{p+3N}'(\xi_2^p), \text{ where } \xi_2^p\epsilon(\tfrac{\nu-1}{n},\tfrac{\nu}{n}).$$

Moreover, from (2.1.2) and (2.3.1) we get

$$f_p[y(\tfrac{\nu-1}{n})] = F_p[y(\tfrac{\nu-1}{n}),y(\tfrac{\nu-1}{n})].$$

Hence

$$\phi_n(\Phi)\Delta_n y(\tfrac{\nu}{n})-\Delta_n^o\Phi y(\tfrac{\nu}{n}) = \begin{cases} 0, \ \nu=0, \\ \frac{1}{n}[y_p'(\xi^p)-y_{p+3N}'(\xi_2^p)], \ \nu=1(1)n, \\ \frac{1}{n}y_{p+3N}''(\xi_1^p)-F_p[y(\tfrac{\nu}{n}),y(\tfrac{\nu-1}{n})] \\ \quad -F_p[y(\tfrac{\nu-1}{n}),y(\tfrac{\nu-1}{n})], \ \nu=1(1)n. \end{cases} \tag{2.3.26}$$

If there exist the constants (2.3.23), then the functions y_p'', y_{p+3N}' and y_{p+3N}'' are bounded. It follows from Lemma 2.1 and from the fact that $y_q \in C^{(1)}[0,1]$ ($q=p$ or $q=p+3N$). We have also

$$F_p[y(\tfrac{\nu}{n}),y(\tfrac{\nu-1}{n})]-F_p[y(\tfrac{\nu-1}{n}),y(\tfrac{\nu-1}{n})] = \sum_{q=1}^{3N} \frac{\partial F_p(\tilde{y})}{\partial y_q}[y_q(\tfrac{\nu}{n})-y_q(\tfrac{\nu-1}{n})]$$
$$= \frac{1}{n}\sum_{q=1}^{3N} \frac{\partial F_p(\tilde{y})}{\partial y_q} y_q'(\xi_3^q),$$

where $\tilde{y}\in(y(\frac{\nu-1}{n}),y(\frac{\nu}{n}))$, $\xi_3^q\in(\frac{\nu-1}{n},\frac{\nu}{n})$. Since $y_q\in C^{(1)}[0,1]$, and from Lemma 2.1 it follows that the functions $\partial F_p/\partial y_q$ are bounded, then

$$F_p[y(\tfrac{\nu}{n}),y(\tfrac{\nu-1}{n})]-F_p[y(\tfrac{\nu-1}{n}),y(\tfrac{\nu-1}{n})] = O(\tfrac{1}{n}).$$

From (2.3.26) and the above considerations we obtain

$$\|\ \phi_n(\Phi)\Delta_n y(\tfrac{\nu}{n}) - \Delta_n^o \Phi y(\tfrac{\nu}{n})\ \| = O(\tfrac{1}{n}),$$

which ought to be proved.

For futher simplicity let us introduce the following notations

$$[[\eta(\tfrac{\nu}{n})]]_1 = \sum_{p=1}^{3N} |\eta_p(\tfrac{\nu}{n})|, \quad [[\eta(\tfrac{\nu}{n})]]_2 = \sum_{p=1}^{3N} |\eta_{p+3N}(\tfrac{\nu}{n})|,$$
$$[[\ .\ ]] = [[\ .\ ]]_1 + [[\ .\ ]]_2. \tag{2.3.27}$$

<u>Theorem 2.6.</u> Let us assume the existence of constants (2.3.23). For $n>3N\tilde{M}$, where $\tilde{M}$ denotes the constant occuring in Lemma 2.2, the method of discrete mechanics (2.3.19) is stable on the initial value problem (2.3.18) with

$$S = e^{3N\tilde{M}}\left\{\frac{3N([\tilde{M}]-\tilde{M})+1}{3N[\tilde{M}]+1}\right\}^{-3N[\tilde{M}]-1}$$

where S is a constant occuring in Def.1.5 and [.] denotes "the integer part of".

Proof. Let z denote the exact solution of (2.3.18) and let for some nonnegative constant r, $\eta_n^{(i)}$ (i=1,2) fulfils the condition

$$\|\Phi_n\eta_n^{(i)}-\Phi_n\Delta_n z\|_{E_n^o} < r, \text{ where } \Phi_n=\phi_n(\Phi):\ E_n\to E_n^o.$$

From Lemma 2.2 we have

$$|F_p[\eta^{(1)}(\tfrac{\nu}{n}),\eta^{(1)}(\tfrac{\nu-1}{n})]-F_p[\eta^{(2)}(\tfrac{\nu}{n}),\eta^{(2)}(\tfrac{\nu-1}{n})]| \le \tilde{M}\sum_{q=1}^{3N}(|\eta_q^{(1)}(\tfrac{\nu}{n})-\eta_q^{(2)}(\tfrac{\nu}{n})|+|\eta_q^{(1)}(\tfrac{\nu-1}{n})-\eta_q^{(2)}(\tfrac{\nu-1}{n})|); \quad p=1(1)3N. \tag{2.3.28}$$

For futher convenience let us denote

$$\varepsilon_p(\tfrac{\nu}{n}) = \eta_p^{(1)}(\tfrac{\nu}{n}) - \eta_p^{(2)}(\tfrac{\nu}{n}),$$
$$\delta_p(\tfrac{\nu}{n}) = \Phi_n\eta_p^{(1)}(\tfrac{\nu}{n}) - \Phi_n\eta_p^{(2)}(\tfrac{\nu}{n}). \tag{2.3.29}$$

Taking into account the condition $\Phi_n\eta\varepsilon E_n^o$ and (2.3.29), we get

$$\varepsilon_p(\tfrac{\nu}{n}) = \varepsilon_p(\tfrac{\nu-1}{n})+\tfrac{1}{2n}[\varepsilon_{p+3N}(\tfrac{\nu}{n})+\varepsilon_{p+3N}(\tfrac{\nu-1}{n})]+\tfrac{1}{n}\delta_p(\tfrac{\nu}{n}),$$

$$\varepsilon_{p+3N}(\tfrac{\nu}{n}) = \varepsilon_{p+3N}(\tfrac{\nu-1}{n})+\tfrac{1}{n}\{F_p[\eta^{(1)}(\tfrac{\nu}{n}),\eta^{(1)}(\tfrac{\nu-1}{n})] - F_p[\eta^{(2)}(\tfrac{\nu}{n}),\eta^{(2)}(\tfrac{\nu-1}{n})]\}+\tfrac{1}{n}\delta_{p+3N}(\tfrac{\nu}{n}).$$

These equations and (2.3.28) yield

$$|\varepsilon_p(\tfrac{\nu}{n})| \le |\varepsilon_p(\tfrac{\nu-1}{n})|+\tfrac{1}{2n}(|\varepsilon_{p+3N}(\tfrac{\nu}{n})|+|\varepsilon_{p+3N}(\tfrac{\nu-1}{n})|)+\tfrac{1}{n}|\delta_p(\tfrac{\nu}{n})|,$$

$$|\varepsilon_{p+3N}(\tfrac{\nu}{n})| \le |\varepsilon_{p+3N}(\tfrac{\nu-1}{n})|+\tfrac{\tilde{M}}{n}\sum_{q=1}^{3N}(|\varepsilon_q(\tfrac{\nu}{n})|+|\varepsilon_q(\tfrac{\nu-1}{n})|) + \tfrac{1}{n}|\delta_{p+3N}(\tfrac{\nu}{n})|.$$

Summing the above inequalities with respect to p=1(1)3N and using the notations (2.3.27), we obtain

$$[[\varepsilon(\tfrac{\nu}{n})]]_1 \le [[\varepsilon(\tfrac{\nu-1}{n})]]_1+\tfrac{1}{2n}(\,[[\varepsilon(\tfrac{\nu}{n})]]_2+[[\varepsilon(\tfrac{\nu-1}{n})]]_2\,) + \tfrac{1}{n}[[\delta(\tfrac{\nu}{n})]]_1,$$

$$[[\varepsilon(\tfrac{\nu}{n})]]_2 \le [[\varepsilon(\tfrac{\nu-1}{n})]]_2+\tfrac{3N\tilde{M}}{n}(\,[[\varepsilon(\tfrac{\nu}{n})]]_1+[[\varepsilon(\tfrac{\nu-1}{n})]]_1\,) + \tfrac{1}{n}[[\delta(\tfrac{\nu}{n})]]_2.$$

If we add both sides of these inequalities and take into account that from the condition $\tilde{M}\ge 1/3N$ it follows $1/(2n)\le 1/n\le 3N\tilde{M}/n$, then

$$[[\varepsilon(\tfrac{\nu}{n})]] \le [[\varepsilon(\tfrac{\nu-1}{n})]] + \tfrac{1}{n}(\tfrac{1}{2}[[\varepsilon(\tfrac{\nu}{n})]]_2+3N\tilde{M}[[\varepsilon(\tfrac{\nu}{n})]]_1 + \tfrac{1}{2}[[\varepsilon(\tfrac{\nu-1}{n})]]_2+3N\tilde{M}[[\varepsilon(\tfrac{\nu-1}{n})]]_1) + \tfrac{1}{n}[[\delta(\tfrac{\nu}{n})]]$$

$$\leq [[\varepsilon(\tfrac{\nu-1}{n})]] + \frac{3N\tilde{M}}{n}([[\varepsilon(\tfrac{\nu}{n})]] + [[\varepsilon(\tfrac{\nu-1}{n})]]) + \frac{1}{n}[[\delta(\tfrac{\nu}{n})]].$$

Hence

$$(1-\frac{3N\tilde{M}}{n})[[\varepsilon(\tfrac{\nu}{n})]] \leq (1+\frac{3N\tilde{M}}{n})[[\varepsilon(\tfrac{\nu}{n})]] + \frac{1}{n}[[\delta(\tfrac{\nu}{n})]].$$

Since $n>3N\tilde{M}$, then

$$[[\varepsilon(\tfrac{\nu}{n})]] \leq \frac{1}{1-\frac{3N\tilde{M}}{n}}\{(1+\frac{3N\tilde{M}}{n})[[\varepsilon(\tfrac{\nu-1}{n})]] + \frac{1}{n}[[\delta(\tfrac{\nu}{n})]]\}.$$

From the last inequality we have

$$[[\varepsilon(\tfrac{\nu}{n})]] \leq \frac{(1+\frac{3N\tilde{M}}{n})^{\nu}}{(1-\frac{3N\tilde{M}}{n})^{\nu}}[[\varepsilon(0)]] + \frac{1}{n}\sum_{\mu=1}^{\nu}\frac{(1+\frac{3N\tilde{M}}{n})^{\nu-\mu}}{(1-\frac{3NM}{n})^{\nu+1-\mu}}[[\delta(\tfrac{\mu}{n})]]. \tag{2.3.30}$$

Since $1-3N\tilde{M}/n>0$ and $3N\tilde{M}\geq 1$, then $0<1-3N\tilde{M}/n<1$, and so if $\alpha_1\geq\alpha_2$, then $(1-3N\tilde{M}/n)^{\alpha_1} \leq (1-3N\tilde{M}/n)^{\alpha_2}$. Therefore, from (2.3.30) we obtain ($[[\varepsilon(0)]]=[[\delta(0)]]$)

$$[[\varepsilon(\tfrac{\nu}{n})]] \leq \frac{(1+3N\tilde{M}/n)^n}{(1-3N\tilde{M}/n)^n}([[\delta(0)]] + \frac{1}{n}\sum_{\nu=1}^{n}[[\delta(\tfrac{\nu}{n})]])$$
$$\leq (1+3N\tilde{M}/n)^n(1-3N\tilde{M}/n)^{-n}||\delta||_{E_n^o}. \tag{2.3.31}$$

Since $n>3N\tilde{M}$, then the sequence $(1-3N\tilde{M}/n)^{-n}$ is decreasing. Thus, for each $n>3N\tilde{M}$,

$$(1-3N\tilde{M}/n)^{-n} \leq (1-\frac{3N\tilde{M}}{1+3N[\tilde{M}]})^{-3N[\tilde{M}]-1}.$$

Moreover, for each n we have $(1+3N\tilde{M}/n)^n\leq e^{3N\tilde{M}}$. As a consequence of this and (2.3.31) we obtain

$$[[\varepsilon(\tfrac{\nu}{n})]] \leq e^{3N\tilde{M}}\{\frac{3N([\tilde{M}]-\tilde{M})+1}{3N[\tilde{M}]+1}\}^{-3N[\tilde{M}]-1}||\delta||_{E_n^o}.$$

Since the above inequality holds for each $\nu=0(1)n$, then the proof is finished and a constant r, occuring in Def.1.5, is equal to

Theorems 2.4 and 2.6 ensure (if there exist suitable constants) the fulfillment of conditions (ii) and (iii) in Th.1.2 on the convergence of discretization methods. It only needs to be shown that the mappings $\Phi_n=\phi_n(\Phi):E_n\to E_n^o$ are continuous in adequate domains $B_R(\Delta_n z)=\{\eta_n\in E_n: ||\eta_n-\Delta_n z||<R\}$, where R denotes some constant independent of n. A determination of Φ_n

follows from the definition of a discrete mechanics method (2.3.19).

Lemma 2.3. If there exist constants (2.3.23), then the mappings $\phi_n(\Phi)$: $E_n \to E_n^o$ for the problem (2.3.19) are continuous in each point of the domain $B_R(\Delta_n z)$.

Proof. Let $\eta_n \in B_R(\Delta_n z)$ denote an arbitrary point of $B_R(\Delta_n z)$. We will show that for each $\varepsilon>0$ there exists $\delta>0$ such that the condition $||\eta_n-\Delta_n z||<\delta$ implies $||\Phi_n\eta_n-\Phi_n(\Delta_n z)||<\varepsilon$. Let us take into account the difference $\Phi_n\eta(\nu/n)-\Phi_n\bar{\eta}(\nu/n)$, where $\bar{\eta}_n \equiv \Delta_n z$. From (2.3.18) and (2.3.19) we have

$$\Phi_n\eta(\tfrac{\nu}{n})-\Phi_n\bar{\eta}(\tfrac{\nu}{n}) = \begin{cases} \eta_p(0)-\bar{\eta}_p(0), \quad \nu=0, \\ \eta_{p+3N}(0)-\bar{\eta}_{p+3N}(0), \quad \nu=0, \\ \eta_p(\frac{\nu}{n})-\bar{\eta}_p(\frac{\nu}{n})-[\eta_p(\frac{\nu-1}{n})-\bar{\eta}_p(\frac{\nu-1}{n})] \\ \quad -\frac{1}{2n}[\eta_{p+3N}(\frac{\nu}{n})-\bar{\eta}_{p+3N}(\frac{\nu}{n})] \\ \quad +\frac{1}{2n}[\eta_{p+3N}(\frac{\nu-1}{n})-\bar{\eta}_{p+3N}(\frac{\nu-1}{n})], \quad \nu=1(1)n, \\ \eta_{p+3N}(\frac{\nu}{n})-\bar{\eta}_{p+3N}(\frac{\nu}{n})-[\eta_{p+3N}(\frac{\nu-1}{n})-\bar{\eta}_{p+3N}(\frac{\nu-1}{n})] \\ \quad -\frac{1}{n}\{F_p[\eta(\frac{\nu}{n}),\eta(\frac{\nu-1}{n})]-F_p[\bar{\eta}(\frac{\nu}{n}),\bar{\eta}(\frac{\nu-1}{n})]\}, \\ \qquad \nu=1(1)n. \end{cases} \tag{2.3.32}$$

From the definition of norm in E_n we obtain

$$\left.\begin{array}{l} |\eta_p(\frac{\nu}{n})-\bar{\eta}_p(\frac{\nu}{n})| \\ |\eta_{p+3N}(\frac{\nu}{n})-\bar{\eta}_{p+3N}(\frac{\nu}{n})| \end{array}\right\} \le ||\eta-\bar{\eta}|| \quad \text{for all } \nu=0(1)n.$$

If there exist constants (2.3.23), then from Lemma 2.2 we have

$$|F_p[\eta(\tfrac{\nu}{n}),\eta(\tfrac{\nu-1}{n})]-F_p[\bar{\eta}(\tfrac{\nu}{n}),\bar{\eta}(\tfrac{\nu-1}{n})]| \le 6N\tilde{M}||\eta-\bar{\eta}||.$$

Since for each n we get $1\ge 1/n$, $6N\tilde{M}\ge 6N\tilde{M}/n$ and from $\tilde{M}\ge 1/3N$ it follows that $6N\tilde{M}\ge 1$, then, including the above inequalities and (2.3.32) in a reckoning, we obtain

$$|\Phi_n\eta(\tfrac{\nu}{n})-\Phi_n\bar{\eta}(\tfrac{\nu}{n})| \le (2+6N\tilde{M})||\eta-\bar{\eta}||.$$

The last inequality is true for each ν, and so

$$||\Phi_n\eta-\Phi_n\bar{\eta}|| \le (2+6N\tilde{M})||\eta-\bar{\eta}||.$$

This implies that there exist $\delta<\varepsilon/(2+6N\tilde{M})$ and it brings the proof to an end.

As an immediate consequence of the above considerations and Th.1.2, we have

Conclusion 2.1. If there exist constants (2.3.23), then the discrete mechanics method (2.3.19) is convergent on the initial value problem (2.3.18).

2.3.3. PRACTICAL REALIZATION OF GREENSPAN'S METHOD

Difference dynamical equations of Greenspan's discrete mechanics, i.e. the relationships (2.3.1),(2.3.4)-(2.3.5), can be rewritten in the form

$$x_{li}^{k+1} = \frac{\Delta t}{2}(v_{li}^{k+1}+v_{li}^{k})+x_{li}^{k}, \tag{2.3.33}$$

$$v_{li}^{k+1} = v_{li}^{k} - \Delta t G \sum_{\substack{j=1 \\ j\neq i}}^{N} m_j \frac{x_{li}^{k+1}+x_{li}^{k}-x_{lj}^{k+1}-x_{lj}^{k}}{r_{ij}^{k+1}r_{ij}^{k}(r_{ij}^{k+1}+r_{ij}^{k})}; \tag{2.3.34}$$

$$l=1,2,3;\ i=1(1)N;$$

where

$$r_{ij}^{k} = [\sum_{l=1}^{3}(x_{li}^{k}-x_{lj}^{k})^2]^{1/2}.$$

From (2.3.33) we have

$$v_{li}^{k+1} = \frac{2}{\Delta t}(x_{li}^{k+1}-x_{li}^{k})-v_{li}^{k}. \tag{2.3.35}$$

Substituting (2.3.35) in (2.3.34) we get

$$x_{li}^{k+1} = x_{li}^{k}+\Delta t v_{li}^{k} - \frac{(\Delta t)^2}{2}G \sum_{\substack{j=1 \\ j\neq i}}^{N} m_j \frac{x_{li}^{k+1}+x_{li}^{k}-x_{lj}^{k+1}-x_{lj}^{k}}{r_{ij}^{k+1}r_{ij}^{k}(r_{ij}^{k+1}+r_{ij}^{k})}. \tag{2.3.36}$$

So, it appears that a system of 3N nonlinear equations with unknown quantities x_{li}^{k+1} $(l=1,2,3;\ i=1(1)N)$ ought to be solved in each step. To solve this system we usually apply the iteration process (1.3.43), taking a solution at the previous moment as an initial approximation.

Equations (2.3.35) and (2.3.36) point out that the discrete mechanics method is a one-step method. Thus, we can use an automatic step size correction (see Sect.1.3.1 and Fig.2).

Below the algorithm for the discrete mechanics of Greenspan with an automatic step size change is adduced. In this algorithm the input data are:

G - gravitational constant;

t_o - initial moment;

Δt - basic step of the solution;

n - number of steps ($n=(T-t_o)/\Delta t$, where T denotes the final moment);

N - number of bodies (material points);

m_i - masses of bodies ($i=1(1)N$);

x_{li}^o - coordinates of positions at the initial moment ($l=1,2,3$; $i=1(1)N$);

v_{li}^o - components of velocities at the initial moment ($l=1,2,3$; $i=1(1)N$);

$\bar{\varepsilon}$ - admissible discretization error;

ε - precision in an iteration process.

The output data are:

x_{li}^k - coordinates of positions at the moments $t_k=t_o+k\Delta t$ ($l=1,2,3$; $i=1(1)N$),

v_{li}^k - components of velocities at t_k ($l=1,2,3$; $i=1(1)N$); $k=1(1)n$.

<u>Algorithm 9.</u>

1° $H:=\Delta t$;
   $S:=$<u>true</u>;
   $k:=0$;

2° $t:=t_o+k\Delta t$;

3° for $l=1,2,3$ and $i=1(1)N$:

$$\bar{x}_{li}^{k+1}:=x_{li}^{k+1}:=x_{li}^k;$$

$$\bar{v}_{li}^{k+1}:=v_{li}^k;$$

4° for $l=1,2,3$ and $i=1(1)N$:

$$x_{li}^{k+1}:=x_{li}^{k+1}-\frac{x_{li}^{k+1}-x_{li}^k-Hv_{li}^k+\frac{H^2G}{2}\sum_{\substack{j=1\\j\neq i}}^{N} m_j\frac{x_{li}^{k+1}+x_{li}^k-x_{lj}^{k+1}-x_{lj}^k}{r_{ij}^{k+1}r_{ij}^k(r_{ij}^{k+1}+r_{ij}^k)}}{1+\frac{H^2G}{2}\sum_{\substack{j=1\\j\neq i}}^{N}\frac{m_j}{r_{ij}^{k+1}r_{ij}^k(r_{ij}^{k+1}+r_{ij}^k)}(1-A_{lij}^k)},$$

where
$$A_{lij}^k=\frac{(x_{li}^{k+1}+x_{li}^k-x_{lj}^{k+1}-x_{lj}^k)(x_{li}^{k+1}-x_{lj}^{k+1})}{r_{ij}^{k+1}}\left(\frac{1}{r_{ij}^{k+1}}+\frac{1}{r_{ij}^{k+1}+r_{ij}^k}\right),$$

$$r_{ij}^s=\left[\sum_{l=1}^{3}(x_{li}^s-x_{lj}^s)^2\right]^{1/2}\quad (s=k \text{ or } s=k+1);$$

for l=1,2,3 and i=1(1)N:

$v_{li}^{k+1} := \frac{2}{H}(x_{li}^{k+1} - x_{li}^{k}) - v_{li}^{k}$;

5° if for each l=1,2,3 and each i=1(1)N it holds

$|x_{li}^{k+1} - \bar{x}_{li}^{k+1}| < \varepsilon$ and $|v_{li}^{k+1} - \bar{v}_{li}^{k+1}| < \varepsilon$ then go to 6°;

for l=1,2,3 and i=1(1)N:

$\bar{x}_{li}^{k+1} := x_{li}^{k+1}$;

$\bar{v}_{li}^{k+1} := v_{li}^{k+1}$;

go to 4°;

6° if S=<u>true</u> then go to 7°;

$H := H_1$;

S:=<u>true</u>;

printout x_{li}^{k+1} and v_{li}^{k+1} (l=1,2,3; i=1(1)N);

k:=k+1;

if $k \le n-1$ then go to 2°;

stop;

7° F:=<u>true</u>;

for l=1,2,3 and i=1(1)N:

$y_{li} := x_{li}^{k}$;

$w_{li} := v_{li}^{k}$;

8° for l=1,2,3 and i=1(1)N:

$\bar{z}_{li} := z_{li} := y_{li}$;

$\bar{u}_{li} := w_{li}$;

9° for l=1,2,3 and i=1(1)N:

$$z_{li} := z_{li} - \frac{z_{li} - y_{li} - \frac{H}{2} w_{li} + \frac{H^2 G}{8} \sum_{\substack{j=1 \\ j \neq i}}^{N} m_j \frac{z_{li} + y_{li} - z_{lj} - y_{lj}}{\tilde{r}_{ij} r_{ij} (\tilde{r}_{ij} + r_{ij})}}{1 + \frac{H^2 G}{8} \sum_{\substack{j=1 \\ j \neq i}}^{N} \frac{m_j}{\tilde{r}_{ij} r_{ij} (\tilde{r}_{ij} + r_{ij})} (1 - A_{lij})},$$

$$\text{where } A_{lij} = \frac{(z_{li} + y_{li} - z_{lj} - y_{lj})(z_{li} - z_{lj})}{\tilde{r}_{ij}} \left(\frac{1}{\tilde{r}_{ij}} + \frac{1}{\tilde{r}_{ij} + r_{ij}} \right),$$

$$r_{ij} = \left[\sum_{l=1}^{3} (y_{li} - y_{lj})^2 \right]^{1/2}, \quad \tilde{r}_{ij} = \left[\sum_{l=1}^{3} (z_{li} - z_{lj})^2 \right]^{1/2};$$

10° if for each l=1,2,3 and i=1(1)N it holds

$|z_{li} - \bar{z}_{li}| < \varepsilon$ and $|u_{li} - \bar{u}_{li}| < \varepsilon$ then:

if F=<u>false</u> then go to 11°;

F:=<u>false</u>;

for l=1,2,3 and i=1(1)N:

$y_{li} := z_{li}$;

```
            w_li:=u_li;
          go to 8°;
    for l=1,2,3 and i=1(1)N:
       z̄_li:=z_li;
       ū_li:=u_li;
    go to 9°;
11° h̃:=H/[2   max   (|x_li^(k+1)-z_li|,|v_li^(k+1)-u_li|)/ε̄]^(1/2);
            l=1,2,3
            i=1(1)N
    if h̃≤H/3 then:
       H:=2h̃;
       go to 3°;
    if h̃≥H then:
       if t+2H≤t_o+(k+1)Δt then:
          H:=2H;
          go to 3°;
       H:=t_o+(k+1)Δt-t;
       H_1:=H;
       S:=false;
       go to 3°;
    for l=1,2,3 and i=1(1)N:
       x_li^k:=z_li;
       v_li^k:=u_li;
    t:=t+H;
    if t +2h̃<t_o+(k+1)Δt then:
       H:=2h̃;
       go to 3°;
    H:=t_o+(k+1)Δt-t;
    H_1:=2h̃;
    S:=false;
    go to 3°.
```

The application of the above algorithm to our model two-body problem (see Sect.2.1 and Table V) yields the results presented in Table XV. Let us note that in agreement with Ths. 2.1-2.3 and taking no account of rounding errors, the constants of motion have the same values at an arbitrary moment (compare Table V). Moreover, we get a better approximation of the exact solution than using conventional numerical methods of similar order (compare e.g. with the modified Euler method or Fig.12 and Fig.6). Unfortunately, from another point of view, the discrete mechanics method of Greenspan with an automatic step size correction is very consumptive of time. This fact one must take into account when solving the N-body problem for N>>2 (see Fig.13, Table XVI).

Table XV. Solution of the two-body problem by Greenspan's discrete mechanics method ($\Delta t=0.01$, $\varepsilon=0.000000001$)

ν	i	l	x_{li}	v_{li}	characteristics constants of motion
20	1	1	0.309019104371	-5.975646637169	$\alpha=318$, $\beta=71$
		2	0.951057380068	1.941624119813	$\Delta=0.000005\%$
	2	1	0.000002075279	0.000017947148	$c_1=-0.000000000369$
		2	0.000000917772	0.000013039366	$c_2=6.283185307029$
					$c_4=1.000000000049$
					$c_5=0.000000000016$
					$c_9=6.283185307156$
					$c_{10}=-19.73909255396$
60	1	1	-0.809011321527	3.693154114882	$\alpha=309$, $\beta=69$
		2	-0.587772313465	-5.083169257692	$\Delta=0.000024\%$
	2	1	0.000005433152	-0.000011091952	$c_1=-0.000000001493$
		2	0.000013087787	0.000034137503	$c_2=6.283185305457$
					$c_4=1.000000000491$
					$c_5=0.000000000624$
					$c_9=6.283185307146$
					$c_{10}=-19.73909255402$
100	1	1	1.000000000005	-0.000002000176	$\alpha=327$, $\beta=73$
		2	0.000019189085	6.283185307160	$\Delta=0.000032\%$
	2	1	0.000000000000	0.000000000000	$c_1=-0.000000002512$
		2	0.000018870747	0.000000000000	$c_2=6.283185286448$
					$c_4=1.000000001274$
					$c_5=0.000000016384$
					$c_9=6.283185307194$
					$c_{10}=-19.73909255373$

CPU time 18.76 sec

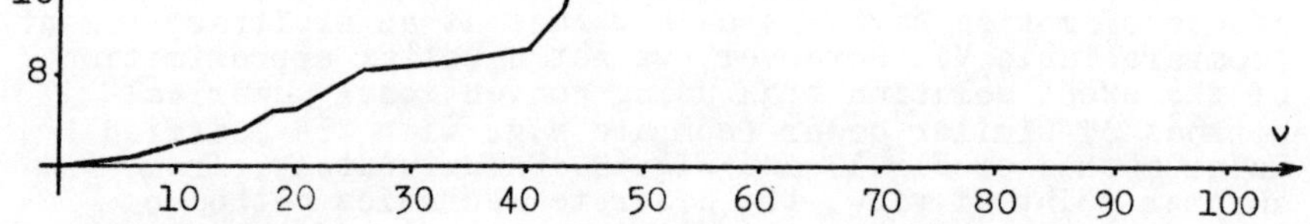

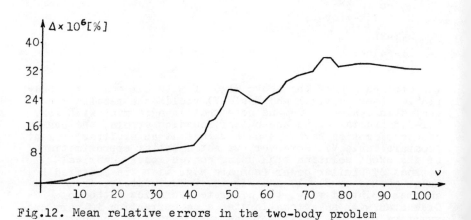

Fig.12. Mean relative errors in the two-body problem

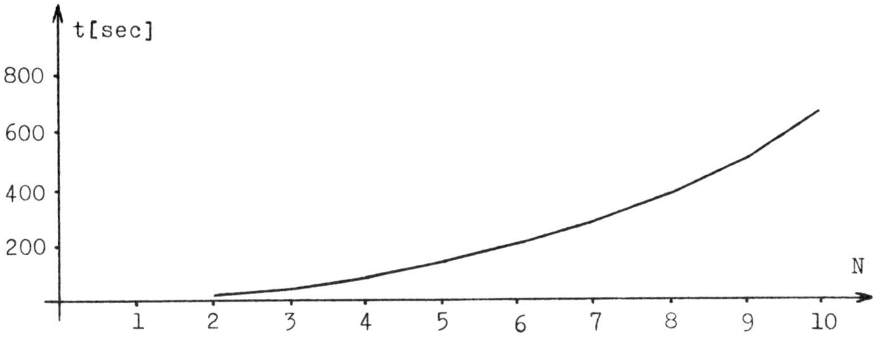

Fig.13. Increase of computational time according to the number of bodies

Table XVI. Solution of the six-body problem by Greenspan's discrete mechanics method (Δt=0.01, ε=0.000000001, the data at the initial moment as in Table XIV)

ν	i	l	x_{li}	v_{li}	characteristics constants of motion
40	1	1	0.7975411885	-3.8460143438	α=309, β=69
		2	0.5915631468	5.0253245654	c_1=-954.55447763
		3	-0.0000009762	-0.0000067519	c_2= 117.70024767
	2	1	0.3168442973	-2.7852629424	c_3= -10.42970171
		2	5.1065151436	0.3001699868	c_4=-541.57467287
		3	-0.0727759155	-0.0490517019	c_5=1451.69367563
	3	1	-7.3653380292	-1.3327795466	c_6= 14.88743782
		2	5.5086186665	-1.6360838091	c_7= 19.83374431
		3	0.1985956976	0.0814069144	c_8= 111.93372063
	4	1	-13.8122787827	0.9483445114	c_9=7394.86717049
		2	-12.4535559641	-1.1334054190	c_{10}=-1447.50653363
		3	0.1330761961	-0.0164965480	
	5	1	-7.5164027113	1.1005934427	
		2	-29.3293864641	-0.2777054433	
		3	0.7762267326	-0.0197752102	
	6	1	0.0000196658	0.0000859747	
		2	0.0001137646	0.0005833923	
		3	-0.0000011369	-0.0000060236	
70	1	1	-0.8347164793	-3.4643680562	α=318, β=71
		2	0.5288848338	-5.3311257830	c_1=-954.55447720
		3	-0.0000044567	-0.0000147174	c_2= 117.70024714
	2	1	-0.5192570802	-2.7765941160	c_3= -10.42970166
		2	5.1291092334	-0.1481687917	c_4=-541.57467311
		3	-0.0864710745	-0.0420761100	c_5=1451.69367592

Table XVI. (cont.)

ν	i	l	x_{li}	v_{li}	characteristics constants of motion
	3	1	-7.7358884746	-1.2180924848	c_6= 14.88743779
		2	5.0056365167	-1.7157396861	c_7= 19.83374394
		3	0.2225464997	0.0782089561	c_8= 111.93372053
	4	1	-13.5239880959	0.9734937127	c_9=7394.86716596
		2	-12.7901065796	-1.1101722543	c_{10}=-1447.50653381
		3	0.1280909544	-0.0167367623	
	5	1	-7.1857508783	1.1037286198	
		2	-29.4108188641	-0.2651711344	
		3	0.7702446118	-0.0201051774	
	6	1	0.0000423212	0.0000425114	
		2	0.0003611183	0.0010635828	
		3	-0.0000037791	-0.0000117408	
100	1	1	-0.2590167488	5.9733035044	α=309, β=69
		2	-0.9803820169	-1.6255002612	c_1=-954.55447646
		3	-0.0000084530	-0.0000122159	c_2= 117.70024662
	2	1	-1.3420260322	-2.6971203661	c_3= -10.42970161
		2	5.0190139879	-0.5825541232	c_4=-541.57467377
		3	-0.0979296766	-0.0341816405	c_5=1451.69367637
	3	1	-8.0957463967	-1.0989343010	c_6= 14.88743775
		2	4.4800213508	-1.7869325879	c_7= 19.83374360
		3	0.2454884734	0.0746850529	c_8= 111.93372047
	4	1	-13.2282429709	0.9980367753	c_9=7394.86716133
		2	-13.1196043061	-1.0863889898	c_{10}=-1447.50653445
		3	0.1230351237	-0.0169671264	
	5	1	-6.8541796567	1.1067226248	
		2	-29.4884855711	-0.2526010906	
		3	0.7641638787	-0.0204326187	
	6	1	0.0000332142	-0.0000969849	
		2	0.0007454474	0.0014857747	
		3	-0.0000082598	-0.0000182426	

CPU time 196.50 sec

2.4. DISCRETE MECHANICS OF ARBITRARY ORDER

The discrete mechanics of Greenspan is a numerical method of the first order (see Th.2.5). Thus, a problem arises to construct a method of higher order for the N-body problem that will be able to conserve the constants of motion. Such a method, as will be shown in this section, can be constructed and consists in some modification of polynomial extrapolation applicable to Greenspan's discrete mechanics.

Let us consider in an inertial frame of reference the motion of an isolated system of N material points with given positions and velocities at the initial moment. Without loss of generality we can assume that we solve our problem in the time interval $[0,1]$. Let us define on this interval a sequence of nets $\{G_k\}_{k=1}^{J}$, where $G_k=1/(\beta_k n)$ and where the step $1/n$ is the basic step of the solution. The sequence $\{\beta_k\}_{k=1}^{J}$ is an arbitrary, exactly increasing sequence of natural numbers. For example, we can choose the Bulirsch sequence given by (1.3.51).

Discrete dynamical equations for the net G_k we define as follows

$$\begin{aligned} 2\beta_{k-1}n(x^{\mu+1}_{(k)li}-x^{\mu}_{(k)li}) &= v^{\mu+1}_{(k)li}+v^{\mu}_{(k)li}, \\ m_i\beta_{k-1}n(v^{\mu+1}_{(k)li}-v^{\mu}_{(k)li}) &= F_{li}(x^{\mu+1}_{(k)},x^{\mu}_{(k)}); \end{aligned} \tag{2.4.1}$$

$$l=1,2,3;\ i=1(1)N;\ \mu=0(1)\beta_{k-1}n-1;$$

where

$$x^{\mu}_{(k)li}\equiv x_{(k)li}\left(\frac{\mu}{\beta_{k-1}n}\right),\quad v^{\mu}_{(k)li}\equiv v_{(k)li}\left(\frac{\mu}{\beta_{k-1}n}\right),$$

$$F_{li}(x^{\mu+1}_{(k)},x^{\mu}_{(k)})=-Gm_i\sum_{\substack{j=1\\ j\neq i}}^{N} m_j\frac{x^{\mu+1}_{(k)li}+x^{\mu}_{(k)li}-x^{\mu+1}_{(k)lj}-x^{\mu}_{(k)lj}}{r^{\mu+1}_{(k)ij}r^{\mu}_{(k)ij}(r^{\mu+1}_{(k)ij}+r^{\mu}_{(k)ij})}. \tag{2.4.2}$$

In (2.4.1) and (2.4.2), $x_{(k)li}$ denotes the l-th coordinate of position ($l=1,2,3$) of the i-th material point with mass m_i ($i=1(1)N$) on the net G_k, and $v_{(k)li}$ is the l-th component of velocity of the i-th body; G denotes, as previously, the gravitational constant, and $r_{(k)ij}$ is the distance between i-th and j-th material points, i.e.

$$r_{(k)ij}=\left[\sum_{l=1}^{3}(x_{(k)li}-x_{(k)lj})^2\right]^{1/2}. \tag{2.4.3}$$

As can be easily noted, the equations (2.4.1)-(2.4.2) are the discrete mechanics equations (compare (2.3.1),(2.3.4)-(2.3.5)) defined on the net G_k. These equations join the positions and velocities of N material points at two succeeding moments $(\mu+1)/(\beta_{k-1}n)$ and $\mu/(\beta_{k-1}n)$ $(\mu=0(1)\beta_{k-1}n-1)$. Moreover, for each net G_k we have $x_{(k)li}(0)=x^o_{li}$ and $v_{(k)li}(0)=v^o_{li}$, where x^o_{li} and v^o_{li} denote initial conditions. Now, let us define the solution at the point ν/n $(\nu=1(1)n)$ as follows

$$x^\nu_{li} = \sum_{k=1}^{J} a_k x^\nu_{(k)li},$$
$$v^\nu_{li} = \sum_{k=1}^{J} a_k v^\nu_{(k)li}; \quad l=1,2,3;\ i=1(1)N; \tag{2.4.4}$$

where the coefficients are determined by the system of equations

$$\sum_{k=1}^{J} a_k = 1,$$
$$\sum_{k=1}^{J} \frac{a_k}{(\beta_{k-1}n)^j} = 0; \quad j=1(1)J-2; \tag{2.4.5}$$
$$\sum_{i=1}^{N} m_i\Big[\sum_{l=1}^{3}\Big(\sum_{k=1}^{J} a_k v^\nu_{(k)li}\Big)^2 - G\sum_{\substack{j=1\\ j\neq i}}^{N}\frac{m_j}{\tilde r^\nu_{ij}}\Big] = 2E^o,$$

where

$$\tilde r^\nu_{ij} = \Big\{\sum_{l=1}^{3}\Big[\sum_{k=1}^{J} a_k\big(x^\nu_{(k)li}-x^\nu_{(k)lj}\big)\Big]^2\Big\}^{1/2}. \tag{2.4.6}$$

E^o denotes the total energy of the system at the initial moment, i.e.

$$E^o = \frac{1}{2}\sum_{i=1}^{N} m_i\Big[\sum_{l=1}^{3}(v^o_{li})^2 - G\sum_{\substack{j=1\\ j\neq i}}^{N}\frac{m_j}{r^o_{ij}}\Big], \tag{2.4.7}$$

where

$$r^o_{ij} = \Big[\sum_{l=1}^{3}(x^o_{li}-x^o_{lj})^2\Big]^{1/2}.$$

From Th.1.8 it is known that if the solutions x^ν_{li} and v^ν_{li} are determined by the formulas (2.4.4), while the coefficients a_k are calculated from the system (2.4.5), in which the last equation is replaced by

$$\sum_{k=1}^{J} \frac{a_k}{(\beta_{k-1}n)^{J-1}} = 0,$$

then these solutions have an accuracy of order J. In this case we have the conventional polynomial extrapolation method. It will be shown further that in our case the solutions of order J-1 are obtained, and that they fulfil most of the laws of mechanics.

Let us note that the last equation of (2.4.5) follows the dependence on the time-moment of a_k, i.e. $a_k=a_k(\nu)$.

The system of equations (2.4.5) is composed of J-1 linear equations and one which is nonlinear. In order to solve the system it is convenient to compute $a_1, a_2, \ldots, a_{J-1}$ from the linear equations and insert these results into the last equation of (2.4.5). In this way we obtain one nonlinear equation with one unknown quantity a_J that can be solved by Newton's iteration process. Such procedure is described below.

Linear equations (2.4.5) can be rewritten in the form

$$\begin{aligned} &\sum_{k=1}^{J} a_k = 1, \\ &\sum_{k=1}^{J} B_k^j a_k = 0; \quad j=1(1)J-2; \end{aligned} \tag{2.4.8}$$

where

$$B_k^j = \Big(\prod_{\substack{l=1 \\ l\neq k}}^{J} \beta_{l-1} \Big)^j .$$

Let us introduce the notations

$$\begin{aligned} &f_{lJ} = \frac{B_J}{B_l} \prod_{\substack{i=1 \\ i\neq l}}^{J-2} \frac{B_J - B_i}{B_l - B_i} , \\ &A = 1 - \sum_{l=1}^{J-2} f_{lJ}, \quad C = 1 - \sum_{l=1}^{J-2} f_{l,J-1}. \end{aligned} \tag{2.4.9}$$

Thus, from (2.4.8) we obtain

$$\begin{aligned} &a_{J-1} = \frac{1}{C} - \frac{A}{C} a_J, \\ &a_l = \Big(\frac{A}{C} f_{l,J-1} - f_{lJ} \Big) a_J - f_{l,J-1}/C; \quad l=1(1)J-2. \end{aligned} \tag{2.4.10}$$

Let

$$\alpha_{l1} = \begin{cases} -f_{l,J-1}/C, \text{ for } l=1(1)J-2, \\ 1/C, \text{ for } l=J-1, \\ 0, \text{ for } l=J, \end{cases}$$

$$\alpha_{2l} = \begin{cases} \frac{A}{C} f_{l,J-1} - f_{lJ}, \text{ for } l=1(1)J-2, \\ -A/C, \text{ for } l=J-1, \\ 1, \text{ for } l=J. \end{cases} \tag{2.4.11}$$

Then

$$a_k = \alpha_{1k}+\alpha_{2k}a_J; \quad k=1(1)J. \tag{2.4.12}$$

Substitution of (2.4.12) in the last equation of (2.4.5) yields

$$\Phi(a_J) \equiv \sum_{i=1}^{N} m_i\{\sum_{l=1}^{3} [\sum_{k=1}^{J} (\alpha_{1k}+\alpha_{2k}a_J)v^{\nu}_{(k)li}]^2 - G \sum_{\substack{j=1\\j\neq i}}^{N} \frac{m_j}{\bar{r}^{\nu}_{ij}} - 2E^{o} = 0 \tag{2.4.13}$$

where

$$\bar{r}^{\nu}_{ij} = \{\sum_{l=1}^{3} [\sum_{k=1}^{J} (\alpha_{1k}+\alpha_{2k}a_J)(x^{\nu}_{(k)li}-x^{\nu}_{(k)lj})]^2\}^{1/2}. \tag{2.4.14}$$

Using Newton's iteration process to solve (2.4.13), we have

$$a_J^{\gamma+1} = a_J^{\gamma} - \frac{\Phi(a_J^{\gamma})}{\frac{\partial\Phi}{\partial a_J}(a_J^{\gamma})}; \quad =0,1,2,\ldots; \tag{2.4.15}$$

where

$$\frac{\partial\Phi}{\partial a_J} = \sum_{i=1}^{N} m_i\{2 \sum_{l=1}^{3} [\sum_{k=1}^{J} (\alpha_{1k}+\alpha_{2k}a_J)v^{\nu}_{(k)li}] \times [\sum_{k=1}^{J} \alpha_{2k}v^{\nu}_{(k)li}] + G \sum_{\substack{j=1\\j\neq i}}^{N} \frac{m_j}{(\bar{r}^{\nu}_{ij})^3} \times \sum_{l=1}^{3} [\sum_{k=1}^{J} (\alpha_{1k}+\alpha_{2k}a_J)(x^{\nu}_{(k)li}-x^{\nu}_{(k)lj})] \times [\sum_{k=1}^{J} \alpha_{2k}(x^{\nu}_{(k)li}-x^{\nu}_{(k)lj})]\}. \tag{2.4.16}$$

As an initial approximation of a_J in the process (2.4.15), it is convenient to take the value obtained from the system (2.4.5) in which the last equation is replaced by

$$\sum_{k=1}^{J} \frac{a_k}{(\beta_{k-1}n)^{J-1}} = 0.$$

It is easy to show that we obtain

$$a_J^o = - \frac{\sum_{k=1}^{J} \frac{\alpha_{1k}}{(\beta_{k-1}n)^{J-1}}}{\sum_{k=1}^{J} \frac{\alpha_{2k}}{(\beta_{k-1}n)^{J-1}}} = - \frac{\sum_{k=1}^{J} \frac{\alpha_{1k}}{\beta_{k-1}^{J-1}}}{\sum_{k=1}^{J} \frac{\alpha_{2k}}{\beta_{k-1}^{J-1}}} . \qquad (2.4.17)$$

Now, let us consider the order problem of the method (2.4.4) -(2.4.5). In [3] it has been shown that the total discretization error for the discrete mechanics formulas (2.4.1)- -(2.4.2) (or (2.3.1),(2.3.4)-(2.3.5)) has an asymptotic expansion of the form

$$\eta^{\mu}_{(k)p} - z^{\mu}_{p} = \sum_{j=1}^{J-1} \frac{1}{(\beta_{k-1}n)^j} e_{pj}\left(\frac{\mu}{\beta_{k-1}n}\right) + O(n^{-J}); \qquad (2.4.18)$$

$$k=1(1)J;\ \mu=0(1)\beta_{k-1}n;\ p=1(1)6N;$$

where $\eta_{(k)p}$ is a solution of (2.4.1) on the net G_k and

$$\eta_{(k)p} = \begin{cases} x_{(k)1i}, & \text{for } p=1(1)3N, \\ v_{(k)1i}, & \text{for } p=3N+1(1)6N. \end{cases}$$

z_p denotes here the exact solution of an adequate continuous problem, i.e. of a differential equation. The functions e_j, as elements of the space $\underset{6N}{\chi} C^{(1)}[0,1]$, are bounded. The relationship (2.4.18) can be deduced from Th.1.3 and from the theorems given in Sect.2.3.2. The existence of constants (2.3.23) must be assumed at the same time.

Using (2.4.18) we can prove

Theorem 2.7. If there exist the constants (2.3.23), then the method (2.4.4)-(2.4.5) of numerical solution of the N-body problem has the order of J-1.

Proof. From the existence of the constants (2.3.23), it follows that the discrete mechanics method of Greenspan is stable (Th.2.6), consistent (Th.2.4) and of the first order (Th.2.5). It means the assumptions (i) and (ii) of Th.1.3 are fulfilled. The fulfillment of other assumptions of this theorem can be proved too ([3]). So, there exists an asymptotic expansion of the form (2.4.18). Let us note that the value J is arbitrary, and hence the expansion (2.4.18) is of an arbitrary order. Let

$$\eta^{\nu}_{p} = \begin{cases} x^{\nu}_{1i}, & \text{for } p=1(1)3N, \\ v^{\nu}_{1i}, & \text{for } p=3N+1(1)6N; \quad \nu=0(1)n; \end{cases}$$

where x_{li}^{ν} and v_{li}^{ν} are given by (2.4.4). To determine the accuracy of the solution η_p, let us estimate the norm

$$||\varepsilon_n|| = ||\eta^{(n)} - z^{(n)}|| = \max_{\nu=0(1)n} \sum_{p=1}^{6N} |\eta_p^{\nu} - z_p^{\nu}|. \quad (2.4.19)$$

From (2.4.4) we have

$$\eta_p^{\nu} = \sum_{k=1}^{J} a_k \eta_{(k)p}^{\nu}; \quad p=1(1)6N.$$

Taking into account (2.4.18), we get

$$\eta_p^{\nu} = \sum_{k=1}^{J} a_k z_p^{\nu} + \sum_{k=1}^{J} a_k \sum_{j=1}^{J-1} \frac{1}{(\beta_{k-1}n)^j} e_{pj}(\tfrac{\nu}{n})$$
$$+ \sum_{k=1}^{J} a_k O(n^{-J})$$
$$= \sum_{k=1}^{J} a_k z_p^{\nu} + \sum_{j=1}^{J-2} \sum_{k=1}^{J} \frac{a_k}{(\beta_{k-1}n)^j} e_{pj}(\tfrac{\nu}{n})$$
$$+ \sum_{k=1}^{J} \frac{a_k}{(\beta_{k-1}n)^{J-1}} e_{p,J-1}(\tfrac{\nu}{n}) + \sum_{k=1}^{J} a_k O(n^{-J});$$
$$\nu=0(1)n.$$

Hence, using (2.4.5), we obtain

$$\eta_p^{\nu} = z_p^{\nu} + \sum_{k=1}^{J} \frac{a_k}{(\beta_{k-1}n)^{J-1}} e_{p,J-1}(\tfrac{\nu}{n}) + O(n^{-J}). \quad (2.4.20)$$

Since $e_{p,J-1} \in C^{(1)}[0,1]$, then they are bounded. Moreover, from (1.3.51) it follows that $1/\beta_{k-1} \leq 1$ for each $k=1(1)J$, so $1/(\beta_{k-1}n)^{J-1} \leq 1$, too. Now we can show that all coefficients a_k $(k=1(1)J)$ are bounded. To do it let us rewrite (2.4.13) in the following form

$$L \equiv \sum_{i=1}^{N} m_i \sum_{l=1}^{3} [\sum_{k=1}^{J} (\alpha_{1k} + \alpha_{2k} a_J) v_{(k)li}^{\nu}]^2$$
$$= G \sum_{i=1}^{N} \sum_{\substack{j=1 \\ j\neq i}}^{N} \frac{m_i m_j}{\bar{r}_{ij}^{\nu}} + 2E^{o} \equiv P, \quad (2.4.21)$$

where $\bar{r}_{ij}^{\nu}$ is given by (2.4.14). While calculating a_J, $x_{(k)li}^{\nu}$ and $v_{(k)li}^{\nu}$ are fixed. For simplicity later on, let us denote

$$V_{ki} \equiv V_{ki}(\nu) = \sum_{l=1}^{3} v_{(k)li}^{\nu}, \quad (2.4.22)$$

$$X_{kij} \equiv X_{kij}(\nu) = \sum_{l=1}^{3} (x^{\nu}_{(k)li} - x^{\nu}_{(k)lj}).$$

Using the inequality $(|a|+|b|+|c|)^2 \leq 3(a^2+b^2+c^2)$ (for arbitrary a,b,c) and (2.4.22), we have

$$\begin{aligned} L &\equiv \sum_{i=1}^{N} \frac{m_i}{3}\Big[\sum_{l=1}^{3} \Big|\sum_{k=1}^{J} (\alpha_{1k}+\alpha_{2k}a_J) v^{\nu}_{(k)li}\Big|\Big]^2 \\ &\geq \sum_{i=1}^{N} \frac{m_i}{3}\Big[\sum_{l=1}^{3} \sum_{k=1}^{J} (\alpha_{1k}+\alpha_{2k}a_J) v^{\nu}_{(k)li}\Big]^2 \\ &= \sum_{i=1}^{N} \frac{m_i}{3}\Big[\sum_{k=1}^{J} (\alpha_{1k}+\alpha_{2k}a_J) V_{ki}\Big]^2 \end{aligned} \tag{2.4.23}$$

and

$$\begin{aligned} P &= G \sum_{i=1}^{N} \sum_{\substack{j=1 \\ j\neq i}}^{N} \frac{m_i m_j}{\sqrt{\sum_{l=1}^{3} \Big[\sum_{k=1}^{J} (\alpha_{1k}+\alpha_{2k}a_J)(x^{\nu}_{(k)li} - x^{\nu}_{(k)lj})\Big]^2}} + 2E^{o} \\ &\leq G \sum_{i=1}^{N} \sum_{\substack{j=1 \\ j\neq i}}^{N} \frac{\sqrt{3}\times m_i m_j}{\Big|\sum_{l=1}^{3} \sum_{k=1}^{J} (\alpha_{1k}+\alpha_{2k}a_J)(x^{\nu}_{(k)li} - x^{\nu}_{(k)lj})\Big|} + 2|E^{o}| \\ &= G \sum_{i=1}^{N} \sum_{j=1}^{N} \frac{\sqrt{3}\times m_i m_j}{\Big|\sum_{k=1}^{J} (\alpha_{1k}+\alpha_{2k}a_J) X_{kij}\Big|} + 2|E^{o}|. \end{aligned} \tag{2.4.24}$$

Thus, from (2.4.21),(2.4.23) and (2.4.24) we get

$$\sum_{i=1}^{N} \frac{m_i}{3}\Big[\sum_{k=1}^{J} (\alpha_{1k}+\alpha_{2k}a_J) V_{ki}\Big]^2 \leq G \sum_{i=1}^{N} \sum_{\substack{j=1 \\ j\neq i}}^{N} \frac{\sqrt{3}\times m_i m_j}{\Big|\sum_{k=1}^{J} (\alpha_{1k}+\alpha_{2k}a_J) X_{kij}\Big|} + 2|E^{o}|. \tag{2.4.25}$$

The above inequality shows that there exists $C=C(\nu)>0$ such that $|a_J|\leq C$. For if it were not like this, i.e. for each $C>0$ it would be $|a_J|>C$, then from (2.4.25) we would obtain a contradiction as $a_J\to\infty$. So, if we take $\tilde{C} = \max_{\nu=1(1)n} C(\nu)$, then all coefficients $a_J=a_J(\nu)$ are bounded by a common constant $\tilde{C}$. Hence, from (2.4.19) and (2.4.20) we obtain $||\varepsilon_n||=O(n^{-J+1})$. This brings the proof to an end.

As we have shown in Sect.2.3.1, the formulas of Greenspan's discrete mechanics have the following properties:

- invariability with respect to translation, rotation and under uniform motion of the frame of reference;
- uniform motion in a straight line of the center of mass;
- conservation of linear and angular momentum;
- conservation of total energy.

The formulas (2.4.4) fulfil most of the above laws of mechanics too.

Theorem 2.8. Let us consider two rectangular frames of reference $x_1x_2x_3$ and $\bar{x}_1\bar{x}_2\bar{x}_3$. If these frames are related by any of the following relations:

(i) $\bar{x}_l=x_l-b_l$ $(l=1,2,3)$, b=const;

(ii) $\bar{x}_p=l_px_1+m_px_2+n_px_3$ $(p=1,2,3)$, where l_p is the direction cosine of the axis $O\bar{x}_p$ with respect to the axis Ox_1, m_p with respect to the axis Ox_2 and n_p with respect to the axis Ox_3;

(iii) $\bar{x}_l=x_l-c_lt$ $(l=1,2,3)$, c=const;

then in the frame $\bar{x}_1\bar{x}_2\bar{x}_3$ the form of the formulas (2.4.4) is invariant.

Proof. We omit the indices l and i in (2.4.4) as unessential in the proof.

Ad.(i). From (2.4.4) and (2.4.5) we have

$$\bar{x}^{\nu} = x^{\nu}-b = \sum_{k=1}^{J} a_k x^{\nu}_{(k)}-b = \sum_{k=1}^{J} a_k x^{\nu}_{(k)} - \sum_{k=1}^{J} a_k b$$

$$= \sum_{k=1}^{J} a_k (x^{\nu}_{(k)}-b) = \sum_{k=1}^{J} a_k \bar{x}^{\nu}_{(k)}.$$

Since $\bar{x}^{\nu}_{(k)}=x^{\nu}_{(k)}-b$ implies ([40]) $\bar{v}^{\nu}_{(k)}=v^{\nu}_{(k)}$, then

$$v^{\nu} = \sum_{k=1}^{J} a_k v^{\nu}_{(k)} = \sum_{k=1}^{J} a_k \bar{v}^{\nu}_{(k)} = \bar{v}^{\nu}.$$

Ad.(ii). We have

$$\bar{x}^{\nu}_p = l_px^{\nu}_1+m_px^{\nu}_2+n_px^{\nu}_3 = l_p \sum_{k=1}^{J} a_k x^{\nu}_{(k)1} + m_p \sum_{k=1}^{J} a_k x^{\nu}_{(k)2}$$

$$+ n_p \sum_{k=1}^{J} a_k x^{\nu}_{(k)3} = \sum_{k=1}^{J} a_k (l_px^{\nu}_{(k)1}+m_px^{\nu}_{(k)2}+n_px^{\nu}_{(k)3})$$

$$= \sum_{k=1}^{J} a_k \bar{x}_{(k)p}; \quad p=1,2,3.$$

Since $\bar{x}^{\nu}_{(k)p}=l_p x^{\nu}_{(k)1}+m_p x^{\nu}_{(k)2}+n_p x^{\nu}_{(k)3}$ implies ([40]) $\bar{v}^{\nu}_{(k)p}=l_p v^{\nu}_{(k)1}+m_p v^{\nu}_{(k)2}+n_p v^{\nu}_{(k)3}$, the proof of $\bar{v}^{\nu}_p=\sum_{k=1}^{J} a_k \bar{v}^{\nu}_{(k)p}$ proceeds analogously as above.

Ad.(iii).

$$\bar{x}^{\nu}= x^{\nu}-ct_{\nu} = \sum_{k=1}^{J} a_k x^{\nu}_{(k)} - \sum_{k=1}^{J} a_k ct_{\nu}$$

$$= \sum_{k=1}^{J} a_k (x^{\nu}_{(k)}-ct_{\nu}) = \sum_{k=1}^{J} a_k \bar{x}^{\nu}_{(k)},$$

where $t_{\nu}=\nu/n$. Since $\bar{x}^{\nu}_{(k)}=x^{\nu}_{(k)}-ct_{\nu}$ implies ([40]) $\bar{v}^{\nu}_{(k)}=v^{\nu}_{(k)}-c$, then $\bar{v}^{\nu}=\sum_{k=1}^{J} a_k v^{\nu}_{(k)}$ and the proof of Th.2.8 is complete.

Theorem 2.9. For the formulas (2.4.4) the following laws hold:

(i) the center of mass of the system moves in a straight line with uniform velocity, i.e.
$M\tilde{x}^{\nu} = c_1 t_{\nu}+c_2$, where $M=\sum_{i=1}^{N} m_i$ and $\tilde{x}=[\tilde{x}_1,\tilde{x}_2,\tilde{x}_3]$ denotes a position of the center of mass,
$$c_1 = \sum_{i=1}^{N} m_i v_i^o, \quad c_2 = \sum_{i=1}^{N} m_i x_i^o, \quad t_{\nu}=\nu/n$$
$(\nu=0(1)n)$;

(ii) the linear momentum of the system does not change, i.e.
$$\sum_{i=1}^{N} m_i v^{\nu} = \sum_{i=1}^{N} m_i v^{\mu}$$
for arbitrary $\nu,\mu=0(1)n$.

Proof. The laws (i) and (ii) are satisfied for (2.4.1) (see [42]).

Ad.(i). From (2.4.4) we have

$$M\tilde{x}^{\nu} = M \sum_{k=1}^{J} a_k \tilde{x}^{\nu}_{(k)} = \sum_{k=1}^{J} a_k M\tilde{x}^{\nu}_{(k)} = \sum_{k=1}^{J} a_k (c_1 t_{\nu}+c_2).$$

Hence, with respect to (2.4.5), we obtain

$$M\tilde{x}^{\nu} = c_1 t_{\nu}+c_2.$$

Ad.(ii). From the second equation of (2.4.4) and from (2.4.5) we get

$$\sum_{i=1}^{N} m_i v_i^{\nu} = \sum_{i=1}^{N} m_i \sum_{k=1}^{J} a_k v_{(k)i}^{\nu} = \sum_{k=1}^{J} a_k \sum_{i=1}^{N} m_i v_{(k)i}^{\nu}$$
$$= \sum_{k=1}^{J} a_k \sum_{i=1}^{N} m_i v_{(k)i}^{\mu} = \sum_{i=1}^{N} m_i \sum_{k=1}^{J} a_k v_{(k)i}^{\mu}$$
$$= \sum_{i=1}^{N} m_i v_i^{\mu}.$$

Theorem 2.10. If the position and velocity of every material point of the isolated system are defined by (2.4.4), then the law of conservation of the total energy of this system holds exatly, i.e. for arbitrary $\mu,\nu=0(1)n$

$$E(\tfrac{\mu}{n}) \equiv K(\tfrac{\mu}{n})+V(\tfrac{\mu}{n}) = K(\tfrac{\nu}{n})+V(\tfrac{\nu}{n}) \equiv E(\tfrac{\nu}{n}), \qquad (2.4.26)$$

where

$$K(\tfrac{\nu}{n}) = \frac{1}{2} \sum_{i=1}^{N} m_i \sum_{l=1}^{3} (v_{li}^{\nu})^2,$$

$$V(\tfrac{\nu}{n}) = -\frac{1}{2} G \sum_{i=1}^{N} \sum_{\substack{j=1 \\ j\neq i}}^{N} \frac{m_i m_j}{r_{ij}^{\nu}}$$

with

$$r_{ij}^{\nu} = [\sum_{l=1}^{3} (x_{li}^{\nu}-x_{lj}^{\nu})^2]^{1/2}.$$

Proof. Because of the last equation in (2.4.5), it is self-evident that $E(\nu/n)=E^0$, where E^0 denotes the total energy of the system at the initial moment, so the equality (2.4.26) holds.

With respect to the above theorems, the method of numerical solution of the N-body problem described in this section can be called a discrete mechanics of arbitrary order (it ought to be restated here that the quantity J in Th.2.7 is arbitrary).

Remark. For the formulas (2.4.4) the principle of conservation of the angular momentum fails, if it is defined by

$$L(\tfrac{\nu}{n}) = \sum_{i=1}^{N} m_i (x_i^{\nu} \times v_i^{\nu}).$$

But if we accept that

$$\tilde{L}(\tfrac{\nu}{n}) = \sum_{k=1}^{J} a_k L_{(k)}(\tfrac{\nu}{n}),$$

then, taking into account the fact ([42]) $L_{(k)}(\frac{\nu}{n})=$

$L_{(k)}(\frac{\mu}{n})$ (for arbitrary $\mu,\nu=0(1)n$) and (2.4.5), we observe that $\tilde{L}(\frac{\mu}{n})=\tilde{L}(\frac{\nu}{n})$.

Now, let us consider the practical point of the realization of our method. At first, let us discuss a problem of choice of the method-order. Since the last equation in (2.4.5) is nonlinear, we are not able to find any recurence relation between solutions obtained with the different values of J. One ought to be reminded here that the method is of the J-1st order. In this connection, the usage of arbitrary order discrete mechanics formulas as a method with order--choice would necessitate solving the system of equations (2.4.5) for each J. The expenditure of calculations would be too high in this case. Therefore, we propose to fix the value of J. Practical calculations show that J should not be greater than 8. For J=8 our method is of order 7. Moreover, from practice it follows that the accuracy ε_1 in an iteration process of calculation a_J does not have to be too high. It is sufficient to take the value of ε_1 between 10^{-5} and 10^{-7}. From another point of view the presented method is a one-step method. Thus, it is possible to apply the procedure described in Sect.1.3.1 for an automatic step size correction.

Below we present the algorithm for the discrete mechanics of order p=J-1 with a constant step size. In this algorithm we have the following input data:

p - desired order of the method;
G - gravitational constant;
t_o - initial moment;
Δt - step of the solution;
n - number of steps ($n=(T-t_o)/\Delta t$, where T denotes the final moment);
N - number of bodies (material points);
m_i - masses of bodies ($i=1(1)N$);
x^o_{li} - coordinates of position at the initial moment ($l=1,2,3$; $i=1(1)N$);
v^o_{li} - components of velocity at the initial moment ($l=1,2,3$; $i=1(1)N$);
ε - accuracy in an iteration process of calculation $x^{\mu+1}_{(k)li}$ (see (2.4.1)-(2.4.2));
ε_1 - accuracy in an iteration process of calculation $a_J = a_{p+1}$ (see (2.4.15)).

The algorithm gives the following output data:

x^ν_{li} - coordinates of position at the moment $t_\nu=t_o+\nu\Delta t$ ($l=1,2,3$; $i=1(1)N$);
v^ν_{li} - components of velocity at t_ν ($l=1,2,3$; $i=1(1)N$); $\nu=1(1)n$.

Algorithm 10

1° $H:=\Delta t$;

for $k=1(1)p$:

compute β_{k-1} from (1.3.51);

$$E^o := \frac{1}{2}\sum_{i=1}^{N} m_i\Big\{\sum_{l=1}^{3}(v^o_{li})^2 - G\sum_{\substack{j=1\\ j\neq i}}^{N}\frac{m_j}{[\sum_{l=1}^{3}(x^o_{li}-x^o_{lj})^2]^{1/2}}\Big\};$$

for $k=1(1)p+1$:

$$B_k := \prod_{\substack{l=1\\ l\neq k}}^{p+1}\beta_{l-1};$$

for $l=1(1)p-1$:

$$F_l := \frac{B_p}{B_l}\prod_{\substack{i=1\\ i\neq l}}^{p-1}\frac{B_p-B_i}{B_l-B_i};$$

$$F'_l := \frac{B_{p+1}}{B_l}\prod_{\substack{i=1\\ i\neq l}}^{p-1}\frac{B_{p+1}-B_i}{B_l-B_i};$$

$$A:=1-\sum_{l=1}^{p-1}F'_l;$$

$$C:=1-\sum_{l=1}^{p-1}F_l;$$

$\alpha_{1,p+1}:=0$;

$\alpha_{2,p+1}:=1$;

$\alpha_{1,p}:=\frac{1}{C}$;

$\alpha_{2,p}:=-\frac{A}{C}$;

for $l=1(1)p-1$:

$\alpha_{1l}:=-\frac{F_l}{C}$;

$\alpha_{2l}:=\frac{A}{C}F_l-F'_l$;

$$a_{p+1}:=\bar{a}_{p+1}:=-\frac{\sum_{k=1}^{p+1}\frac{\alpha_{1k}}{\beta^p_{k-1}}}{\sum_{k=1}^{p+1}\frac{\alpha_{2k}}{\beta^p_{k-1}}};$$

$\nu:=0$;

2° $t:=t_o+\nu\Delta t$;

$k:=1$;

3° for $l=1,2,3$ and $i=1(1)N$:

$z_{li} := x_{li}$;

$u_{li} := v_{li}$;

$\mu := 0$;

4° for $l=1,2,3$ and $i=1(1)N$:

$\bar{y}_{li} := y_{li} := z_{li}$;

$\bar{w}_{li} := u_{li}$;

5° for $l=1,2,3$ and $i=1(1)N$:

$$y_{li} := y_{li} - \frac{y_{li} - z_{li} - \frac{H}{\beta_{k-1}} u_{li} + \frac{H^2 G}{2\beta_{k-1}^2} \sum_{\substack{j=1\\ j\neq i}}^{N} m_j \frac{y_{li}+z_{li}-y_{lj}-z_{lj}}{\bar{r}_{ij} r_{ij} (\bar{r}_{ij}+r_{ij})}}{1 + \frac{H^2 G}{2\beta_{k-1}^2} \sum_{\substack{j=1\\ j\neq i}}^{N} \frac{m_j}{\bar{r}_{ij} r_{ij} (\bar{r}_{ij}+r_{ij})} (1 - A_{lij})},$$

where $A_{lij} = \frac{(y_{li}+z_{li}-y_{lj}-z_{lj})(y_{li}-y_{lj})}{\bar{r}_{ij}} \left(\frac{1}{\bar{r}_{ij}} + \frac{1}{\bar{r}_{ij}+r_{ij}}\right)$,

$$\bar{r}_{ij} = \left[\sum_{l=1}^{3} (y_{li}-y_{lj})^2\right]^{1/2}, \quad r_{ij} = \left[\sum_{l=1}^{3} (z_{li}-z_{lj})^2\right]^{1/2};$$

for $l=1,2,3$ and $i=1(1)N$:

$w_{li} := \frac{2\beta_{k-1}}{H}(y_{li}-z_{li}) - u_{li}$;

if for each $l=1,2,3$ and $i=1(1)N$ it holds

$|y_{li}-\bar{y}_{li}| < \varepsilon$ and $|w_{li}-\bar{w}_{li}| < \varepsilon$ then:

go to 6°;

for $l=1,2,3$ and $i=1(1)N$:

$\bar{y}_{li} := y_{li}$;

$\bar{w}_{li} := w_{li}$;

go to 5°;

6° $\mu := \mu+1$;

if $\mu \leq \beta_{k-1}-1$ then:

for $l=1,2,3$ and $i=1(1)N$:

$z_{li} := y_{li}$;

$u_{li} := w_{li}$;

go to 4°;

for $l=1,2,3$ and $i=1(1)N$:

$x_{(k)li} := y_{li}$;

$v_{(k)li} := w_{li}$;

$k := k+1$;

if $k \leq p+3$ then:

go to 3°;

7° $f := \sum_{i=1}^{N} m_i \{ \sum_{l=1}^{3} [\sum_{k=1}^{p+1} (\alpha_{1k}+\alpha_{2k}a_{p+1}) v_{(k)li}]^2$

$$-G \sum_{\substack{j=1\\j\neq i}}^{N} \frac{m_j}{\{\sum_{l=1}^{3}[\sum_{k=1}^{p+1}(\alpha_{1k}+\alpha_{2k}a_{p+1})(x_{(k)li}-x_{(k)lj})]^2\}^{1/2}}\}$$

$$f' := \sum_{i=1}^{N} m_i \{2 \sum_{l=1}^{3} [\sum_{k=1}^{p+1}(\alpha_{1k}+\alpha_{2k}a_{p+1}) v_{(k)li}][\sum_{k=1}^{p+1} \alpha_{2k} v_{(k)li}]$$

$$+G \sum_{\substack{j=1\\j\neq i}}^{N} \frac{m_j \sum_{l=1}^{3}\sum_{k=1}^{p+1}(\alpha_{1k}+\alpha_{2k}a_{p+1})(x_{(k)li}-x_{(k)lj})}{\{\sum_{l=1}^{3}[\sum_{k=1}^{p+1}(\alpha_{1k}+\alpha_{2k}a_{p+1})(x_{(k)li}-x_{(k)lj})]^2\}^{3/2}}$$

$$\times[\sum_{k=1}^{p+1} \alpha_{2k}(x_{(k)li}-x_{(k)lj})]\};$$

$a_{p+1} := a_{p+1} - \frac{f-2E^o}{f'}$;

if $|a_{p+1}-\bar{a}_{p+1}| \geq \varepsilon_1$ then:

$\bar{a}_{p+1} := a_{p+1}$;

go to 7°;

8° for k=1(1)p:

compute a_k from (2.4.12);

for l=1,2,3 and i=1(1)N:

$x_{li}^{\nu+1} := \sum_{k=1}^{p+1} a_k x_{(k)li}$;

$v_{li}^{\nu+1} := \sum_{k=1}^{p+1} a_k v_{(k)li}$;

printout $x_{li}^{\nu+1}$ and $v_{li}^{\nu+1}$;

$\nu := \nu+1$;

if $\nu \leq n-1$ then:

go to 2°.

The application of this algorithm to our model two-body problem (see Table V) yields the results presented in Tables XVII and XVIII. Let us note that omitting round-off errors, the constants of motion have the same values as at the initial moment (compare Table V). Moreover, let us pay our attention to the fact that the extrapolation coefficients a_k (k=1(1)J) are not constant. They oscillate round some mean values and at each moment we have (see also Fig.15)

$$\sum_{k=1}^{p+1} a_k = 1 \pm \varepsilon_1.$$

Table XVII. Solution of the two-body problem by the discrete mechanics of order 4 (ε=0.000000000001, ε_1=0.000001, Δt=0.01)

ν	i	l	x_{li}	v_{li}	characteristics(*) coefficients constants of motion
20	1	1	0.309019022951	-5.975646803242	α=64 , β=16
		2	0.951057406545	1.941623608344	Δ=0.000009%
	2	1	0.000002075280	0.000017947149	a_1= 0.033276
		2	0.000000917772	0.000013039367	a_2= -1.998281
					a_3= 13.492262
					a_4=-21.324163
					a_5= 10.796905
					c_1=-0.000000000015
					c_2= 6.283185307173
					c_4= 1.000000000002
					c_5= 0.000000000001
					c_9= 6.283185307180
					c_{10}=-19.73909255381
60	1	1	-0.809011174159	3.693155389613	α=65 , β=16
		2	-0.587772516383	-5.083168331212	Δ=0.000049%
	2	1	0.000005433151	-0.000011091956	a_1= 0.033485
		2	0.000013087788	0.000034137500	a_2= -2.004558
					a_3= 13.520509
					a_4=-21.357640
					a_5= 10.808204
					c_9 and c_{10} as above
100	1	1	1.000000000001	-0.000004640117	α=58 , β=16
		2	0.000019609246	6.283185307173	Δ=0.000074%
	2	1	0.000000000000	0.000000000013	a_1= 0.033296
		2	0.000018870745	0.000000000000	a_2= -1.998884
					a_3= 13.494977
					a_4=-21.327380
					a_5= 10.797991
					c_9 and c_{10} as above
					CPU time 24.19 sec

(*) Explanation of notations - see p.69

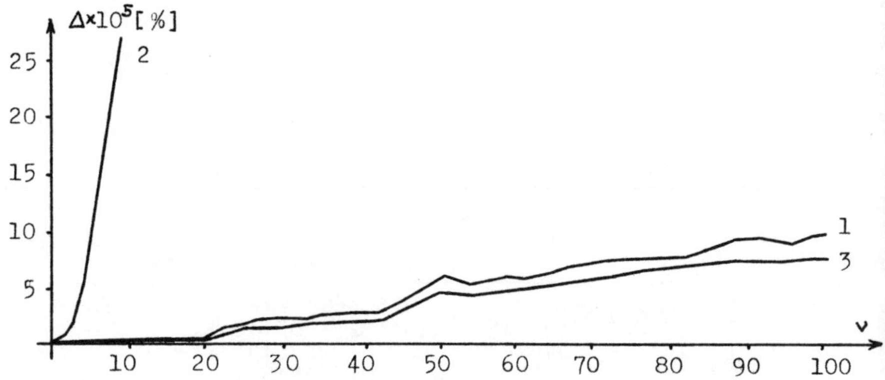

Fig.14. Comparison of accuracy for the methods of order 4: 1 - the Runge-Kutta method with an automatic step size correction, 2 - the method of Adams-Bashforth with constant step, 3 - discrete mechanics with constant step

Table XVIII. Solution of the two-body problem by the discrete mechanics of order 3, 4 and 5 (ε=0.0000000001, ε_1=0.000001, Δt=0.05)

order	ν	i	l	x_{li}	v_{li}	characteristics coefficients
3	10	1	1	-0.9999938757	0.0000008014	α=40, β=10
			2	0.0000093078	-6.2831483042	Δ=0.000025%
		2	1	0.0000060067	0.0000000000	a_1= -0.08968247
			2	0.0000094354	0.0000377415	a_2= 3.07618959
						a_3=-11.42142657
						a_4= 9.43491945
	20	1	1	1.0000000000	-0.0000016029	α=40, β=10
			2	0.0000191259	6.2831853072	Δ=0.000026%
		2	1	0.0000000000	0.0000000000	a_1= -0.08968247
			2	0.0000188707	0.0000000000	a_2= 3.07618960
						a_3=-11.42142660
						a_4= 9.43491947
4	10	1	1	-0.9999938757	0.0000027913	α=63, β=16
			2	0.0000089911	-6.2831483042	Δ=0.000056%
		2	1	0.0000060067	0.0000000000	a_1= 0.03567859
			2	0.0000094354	0.0000377415	a_2= -2.07035760
						a_3= 13.81660921
						a_4=-21.70857387
						a_5= 10.92664368
	20	1	1	1.0000000000	-0.0000055826	α=63, β=16
			2	0.0000197592	6.2831853072	Δ=0.000088%
		2	1	0.0000000000	0.0000000000	a_1= 0.03567858
			2	0.0000188707	0.0000000000	a_2= -2.07035753

Table XVIII. (cont.)

order	ν	i	l	x_{li}	v_{li}	characteristics coefficients
						$a_3 = 13.81660888$
						$a_4 = -21.70857349$
						$a_5 = 10.92664355$
5	10	1	1	-0.9999938757	0.0000023204	$\alpha=94$, $\beta=24$
			2	0.0000090661	-6.2831483042	$\Delta=0.000048\%$
		2	1	0.0000060067	0.0000000000	$a_1 = -0.00284356$
			2	0.0000094354	0.0000377415	$a_2 = 0.53238250$
						$a_3 = -7.01229826$
						$a_4 = 19.18478669$
						$a_5 = -30.22459652$
						$a_6 = 18.52256915$
	20	1	1	1.0000000000	-0.0000046408	$\alpha=94$, $\beta=24$
			2	0.0000196094	6.2831853072	$\Delta=0.000074\%$
		2	1	0.0000000000	0.0000000000	$a_1 = -0.00284354$
			2	0.0000188707	0.0000000000	$a_2 = 0.53238087$
						$a_3 = -7.01228502$
						$a_4 = 19.18476054$
						$a_5 = -30.22457005$
						$a_6 = 18.52255720$

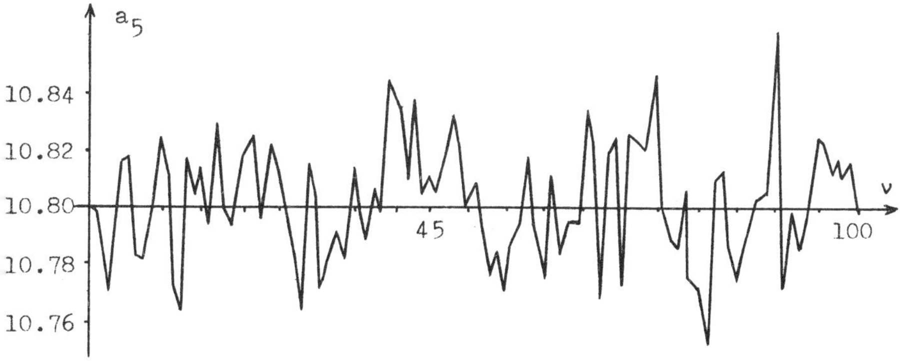

Fig.15. The oscillation of a_5 for the discrete mechanics of order 4 ($\Delta t=0.01$)

Comparing the accuracy of the solutions obtained by conventional and discrete mechanics methods of the same order, we see that discrete mechanics yields a better approximation (see Fig.14). Moreover, discrete mechanics conserves constants of motion.

In Table XIX we present an example of the application of fourth-order discrete mechanics for solving the six-body problem. Let us note that most of the constants of motion, including especially the total energy, are really numerically stable, while even the rational extrapolation method does not have this property (compare Table XIV).

Table XIX. Solution of the six-body problem by the discrete mechanics of order 4 (ε=0.00000001, ε_1=0.00001, Δt=0.05, the data at the initial moment as in Table XIV)

ν	i	l	x_{li}	v_{li}	characteristics coefficients constants of motion
8	1	1	0.79754106	-3.84601543	α=65, β=16
		2	0.59156333	5.02532369	a_1= 0.035683
		3	-0.00000098	-0.00000675	a_2= -2.070483
	2	1	0.31684430	-2.78526294	a_3= 13.817172
		2	5.10651514	0.30016999	a_4=-21.709241
		3	-0.07277592	-0.04905170	a_5= 10.926869
	3	1	-7.36533803	-1.33277955	c_1= -954.554477572
		2	5.50861867	-1.63608380	c_2= 117.700248202
		3	0.19859567	0.08140690	c_3= -10.429703622
	4	1	-13.81227878	0.94834451	c_4= -541.574672823
		2	-12.45355596	-1.13340542	c_5= 1451.693675512
		3	0.13307620	-0.01649655	c_6= 14.887435923
	5	1	-7.51640271	1.10059344	c_7= 19.833730446
		2	-29.32938648	-0.27770551	c_8= 111.933710616
		3	0.77622673	-0.01977521	c_9= 7394.867175400
	6	1	0.00001967	0.00008597	c_{10}=-1447.506534332
		2	0.00011376	0.00058339	
		3	-0.00000114	-0.00000602	
14	1	1	-0.83471668	-3.46436598	α=65, β=16
		2	0.52888451	-5.33112716	a_1= 0.035702
		3	-0.00000446	-0.00001472	a_2= -2.071072
	2	1	-0.51925708	-2.77659411	a_3= 13.819826
		2	5.12910924	-0.14816878	a_4=-21.712386
		3	-0.08647107	-0.04207611	a_5= 10.927930
	3	1	-7.74808890	-1.21809249	c_1= -954.554477571
		2	5.00563652	-1.71573968	c_2= 117.700248202
		3	0.22254647	0.07820894	c_3= -10.429703622
	4	1	-13.52398810	0.97349370	c_4= -541.574672823
		2	-12.79010658	-1.11017226	c_5= 1451.693675512
		3	0.12809095	-0.01673676	c_6= 14.887435923

Table XIX. (cont.)

ν	i	l	x_{li}	v_{li}	characteristics coefficients constants of motion
	5	1	-7.18575088	1.10372861	$c_7 =$ 19.833730445
		2	-29.41081891	-0.26517126	$c_8 =$ 111.933710616
		3	0.77024461	-0.02010518	$c_9 =$ 7394.867175390
	6	1	0.00004232	0.00004251	$c_{10} =$ -1447.506534332
		2	0.00036112	0.00106358	
		3	-0.00000378	-0.00001174	
20	1	1	-0.25901616	5.97330452	$\alpha=65$, $\beta=16$
		2	-0.98038216	-1.62549678	$a_1 =$ 0.035669
		3	-0.00000845	-0.00001222	$a_2 =$ -2.070080
	2	1	-1.34202603	-2.69712037	$a_3 =$ 13.815358
		2	5.01901399	-0.58255411	$a_4 =$ -21.707091
		3	-0.09792968	-0.03418164	$a_5 =$ 10.926143
	3	1	-8.09574640	-1.09893431	$c_1 =$ -954.554477570
		2	4.48002136	-1.78693258	$c_2 =$ 117.700248201
		3	0.24548843	0.07468504	$c_3 =$ -10.429703622
	4	1	-13.22824298	0.99803876	$c_4 =$ -541.574672825
		2	-13.11960431	-1.08638900	$c_5 =$ 1451.693675513
		3	0.12303512	-0.01696713	$c_6 =$ 14.887435923
	5	1	-6.85417966	1.10672261	$c_7 =$ 19.833730445
		2	-29.48848566	-0.25260127	$c_8 =$ 111.933710616
		3	0.76416388	-0.02043262	$c_9 =$ 7394.867175380
	6	1	0.00003521	-0.00009698	$c_{10} =$ -1447.506534332
		2	0.00074545	0.00148577	
		3	-0.00000826	-0.00001824	
					CPU time 8.52 sec

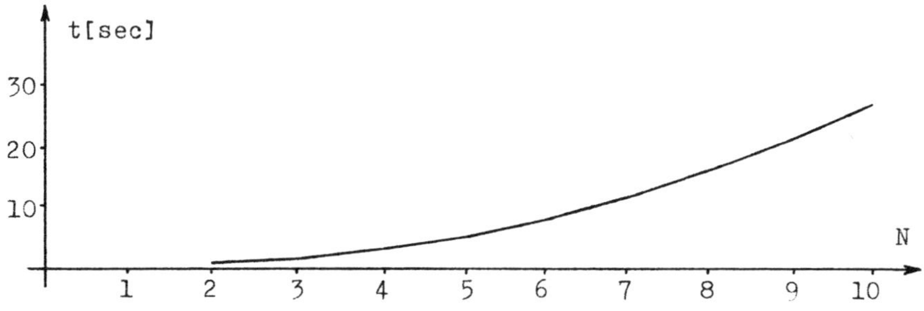

Fig.16. Increase of computational time with respect to the number of bodies for the discrete mechanics of order 4 ($\Delta t=0.05$)

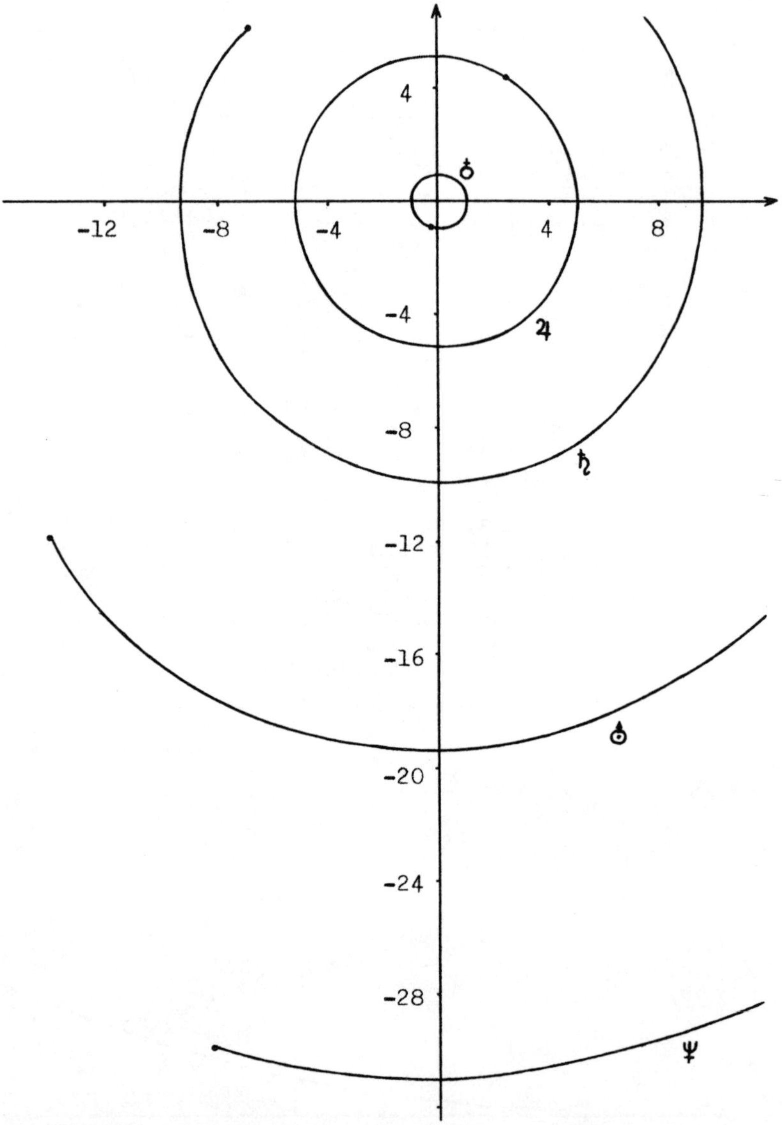

Fig.17. The orbits of Earth, Jupiter, Saturn, Uranus and Neptune (the six-body problem)

2.5. ENERGY CONSERVING METHODS

Discrete mechanics of arbitrary order, described in the previous section, suggests applying the modification (2.4.5) of polynomial extrapolation to an arbitrary method of some order. In this way we would obtain a numerical solution of the N-body problem that would have a high order of accuracy and conserve the total energy. The conservation of the remaining constants would depend on the method used in such an algorithm. Since the constancy of the total energy for an isolated system of N bodies plays a central role in mechanics, we will not be further engaged in the remaining constants of motion.

Thus, let us assume that for $t\in[0,1]$, we solve the initial value problem (2.1.1)-(2.1.2) by an arbitrary one-step method of the form

$$\begin{aligned} x^{o}_{li} &= x_{li}(0), \quad v^{o}_{li} = v_{li}(0), \\ x^{\nu+1}_{li} &= x^{\nu}_{li} + \frac{1}{n}\,\Phi_{li}(x^{\nu+1},x^{\nu},v^{\nu+1},v^{\nu}), \\ v^{\nu+1}_{li} &= v^{\nu}_{li} + \frac{1}{n}\,\Psi_{li}(x^{\nu+1},x^{\nu},v^{\nu+1},v^{\nu}); \\ & l=1,2,3;\ i=1(1)N;\ \nu=0(1)n-1. \end{aligned} \tag{2.5.1}$$

For appropriate expressions Φ_{li} and Ψ_{li}, well-known methods are obtained. For instance, we have:

- Euler's method (see (1.3.3))

$$\Phi_{li} = v^{\nu}_{li}, \quad \Psi_{li} = F_{li}(\tfrac{\nu}{n})/m_i,$$

  where F_{li} is given by (2.1.2);

- discrete mechanics of Greenspan (see (2.3.4)-(2.3.5))

$$\Phi_{li} = \frac{1}{2}(v^{\nu+1}_{li}+v^{\nu}_{li}),$$

$$\Psi_{li} = -G \sum_{\substack{j=1\\ j\neq i}}^{N} m_j \frac{x^{\nu+1}_{li}+x^{\nu}_{li}-x^{\nu+1}_{lj}-x^{\nu}_{lj}}{r^{\nu+1}_{ij}r^{\nu}_{ij}(r^{\nu+1}_{ij}+r^{\nu}_{ij})};$$

- fourth-order method of Runge-Kutta

$$\begin{aligned} \Phi_{li} &= v^{\nu}_{li} + \frac{1}{6n}(A^{(1)}_{li}+A^{(2)}_{li}+A^{(3)}_{li}), \\ \Psi_{li} &= \frac{1}{6}(A^{(1)}_{li}+2A^{(2)}_{li}+2A^{(3)}_{li}+A^{(4)}_{li}), \end{aligned} \tag{2.5.2}$$

  where

$$A^{(1)}_{li} = F_{li}(\tfrac{\nu}{n})/m_i,$$

$$A_{li}^{(2)}=F_{li}(x^\nu+\frac{1}{2n}v^\nu)/m_i,$$
$$A_{li}^{(3)}=F_{li}(x^\nu+\frac{1}{2n}v^\nu+\frac{1}{4n^2}A^{(1)})/m_i,$$
$$A_{li}^{(4)}=F_{li}(x^\nu+\frac{1}{n}v^\nu+\frac{1}{2n^2}A^{(2)})/m_i,$$

and where F_{li}, as noted previously, is defined by (2.1.2).

For later simplicity, in place of (2.5.1) we write

$$\begin{aligned}\eta_p^o &= \eta_p(0),\\ \eta_p^{\nu+1} &= \eta_p^\nu + \frac{1}{n}\Omega_p(\eta^{\nu+1},\eta^\nu);\quad p=1(1)6N;\ \nu=0(1)n-1.\end{aligned} \tag{2.5.3}$$

Moreover, it is assumed that the method (2.5.3) has the order P, i.e. that

$$\varepsilon(\tfrac{\nu}{n}) \equiv \zeta(\tfrac{\nu}{n}) - \eta(\tfrac{\nu}{n}) = O(n^{-P}),$$

where ζ denotes the unique exact solution of the initial value problem (2.1.1)-(2.1.2), and this method has an asymptotic expansion of the total discretization error up to the order J-1, i.e. for each $\nu=0(1)n$

$$\varepsilon(\tfrac{\nu}{n}) = \sum_{j=P}^{J-1}\frac{1}{n^j}e_j(\tfrac{\nu}{n}) + O(n^{-J}), \tag{2.5.4}$$

where the functions e_j are independent of n.

An algorithm to construct the solution of order J-1 conserving the total energy is analogous to the discrete mechanics of arbitrary order method (see the previous section). Namely, the solution at the moment ν/n we define as follows

$$\eta_p^\nu = \sum_{k=P}^{J} a_k \eta_{(k)p}^\nu;\quad \nu=1(1)n;\ p=1(1)6N; \tag{2.5.5}$$

where $\eta_{(k)}$ denotes the solution obtained on the net G_k (k=P(1)J) and the coefficients a_k are calculated from the system of equations of the form

$$\begin{aligned}&\sum_{k=P}^{J} a_k = 1,\\ &\sum_{k=P}^{J}\frac{a_k}{(\beta_{k-P}n)^j} = 0;\quad j=P(1)J-2;\\ &\sum_{i=1}^{N} m_i\Big[\sum_{l=1}^{3}\Big(\sum_{k=P}^{J} a_k v_{(k)li}^\nu\Big)^2 - G\sum_{\substack{j=1\\ j\neq i}}^{N}\frac{m_j}{\tilde r_{(k)ij}^\nu}\Big] - 2E^o = 0,\end{aligned} \tag{2.5.6}$$

where

$$\tilde{r}^{\nu}_{(k)ij} = \{ \sum_{l=1}^{3} [\sum_{k=P}^{J} a_k (x^{\nu}_{(k)li} - x^{\nu}_{(k)lj})]^2 \}^{1/2}. \qquad (2.5.7)$$

In (2.5.6), E^{o} denotes the total energy at the initial moment, $v_{(k)li} \equiv \eta_{(k)p+3N}$, $x_{(k)li} \equiv \eta_{(k)p}$ $(p=i+(l-1)N)$ and β_k is given by (1.3.51).

As it is well-known (see e.g.[67],[86]), if the coefficients a_k are calculated from the following system of equations

$$\sum_{k=P}^{J} a_k = 1,$$

$$\sum_{k=P}^{J} \frac{a_k}{(\beta_{k-P} n)^j} = 0; \quad j=P(1)J-1;$$

then the solution η^{ν} defined by (2.5.5) has an accuracy of order J. In this case we have a conventional polynomial extrapolation. Thus, our method can be construed to be some modification of this extrapolation.

For the method (2.5.5)-(2.5.6) we have

Theorem 2.11. If we use a one-step method of order P to solve the N-body problem (2.1.1)-(2.1.2), and there exists an asymptotic expansion of the total discretization error (2.5.4) for this method, then the solution obtained from (2.5.5)-(2.5.6) has an accuracy of order J-1.

The proof of this theorem, as similar to the proof of Th.2.7, we leave to the reader.

In view of the last equation in (2.5.6), it is obvious:

Theorem 2.12. The method (2.5.5)-(2.5.6) for solving the N-body problem conserves the total energy.

In the method (2.5.5)-(2.5.6) it is especially convenient, taking into account the fast attainment of high-order approximation, to use methods for which there exists asymptotic expansion of total discretization error with respect to even powers of 1/n, i.e. for which

$$\varepsilon(\tfrac{\nu}{n}) = \sum_{j=P}^{J-1} \frac{1}{n^{2j}} e_{2j}(\tfrac{\nu}{n}) + O(n^{-2J-1}).$$

The Gragg method (1.3.58) is an example of such methods. For these types of methods the system of equations (2.5.6) has the form

$$\sum_{k=P}^{J} a_k = 1,$$

$$\sum_{k=P}^{J} \frac{a_k}{(\beta_{k-P}n)^{2j}} = 0; \quad j=P(1)J-2;$$

$$\sum_{i=1}^{N} m_i[\sum_{l=1}^{3}(\sum_{k=P}^{J} a_k v^{\nu}_{(k)li})^2 - G\sum_{\substack{j=1\\j\neq i}}^{N} \frac{m_j}{\tilde{r}^{\nu}_{(k)ij}} - 2E^o = 0,$$

where $\tilde{r}^{\nu}_{(k)ij}$ is given by (2.5.7). It is easy to prove that in this case the solution determined from (2.5.5) has the order of accuracy equal to 2J-2 and, of course, this solution conserves the total energy of the system of bodies.

Below we present three algorithms which use the modification of polynomial extrapolation. In these algorithms the modified Euler method, Runge-Kutta method of fourth order and Gragg method are taken as the basic methods. The algorithms give the solutions of order p=J-1, in the first two cases, and of order p=2J-2 for the Gragg method. In each algorithm we have the following data:

p - desired order of the solution (in the third algorithm p is one half of the order);
G - gravitational constant;
t_o - initial moment;
Δt - step size;
n - number of steps ($n=(T-t_o)/\Delta t$, where T denotes a final moment);
N - number of bodies (material points);
m_i - masses of bodies ($i=1(1)N$);
x^o_{li} - coordinates at the initial moment ($l=1,2,3$; $i=1(1)N$);
v^o_{li} - components of velocities at the initial moment ($l=1,2,3$; $i=1(1)N$);
ε - accuracy in an iteration process of calculation $a_J=a_{p+1}$.

As the results we get:

x^{ν}_{li} - coordinates at the moments $t_{\nu}=t_o+\nu\Delta t$ ($l=1,2,3$; $i=1(1)N$);
v^{ν}_{li} - components of velocities at the moments t_{ν} ($l=1,2,3$; $i=1(1)N$); $\nu=1(1)n$.

Algorithm 11. The modified Euler method with energy conserving modification of polynomial extrapolation

1° $H:=\Delta t$;
 for $k=2(1)p+1$:
 compute β_{k-2} from (1.3.51);

$$E^o := \frac{1}{2}\sum_{i=1}^{N} m_i\Big[\sum_{l=1}^{3}(v^o_{li})^2 - G\sum_{\substack{j=1\\ j\neq i}}^{N}\frac{m_j}{[\sum_{l=1}^{3}(x^o_{li}-x^o_{lj})^2]^{1/2}}\Big];$$

$\nu:=0;$

2° for $k=2(1)p+1$:

$$B_k := \prod_{\substack{l=2\\ l\neq k}}^{p+1}\beta_{l-2};$$

for $l=2(1)p-1$ and $s=p,p+1$:

$$f_{ls} := \frac{B_s}{B_l}\prod_{\substack{i=2\\ i\neq l}}^{p-1}\frac{B_s-B_i}{B_l-B_i};$$

$$A:=1-\sum_{l=2}^{p-1} f_{l,p+1};$$

$$C:=1-\sum_{l=2}^{p-1} f_{lp};$$

for $l=2(1)p-1$:

$\alpha_{1l}:=-f_{lp}/C;$

$\alpha_{2l}:=Af_{lp}/C-f_{l,p+1};$

$\alpha_{1p}:=1/C;$

$\alpha_{2p}:=-A/C;$

$\alpha_{1,p+1}:=0;$

$\alpha_{2,p+1}:=1;$

$$\tilde{a}_{p+1}:=a_{p+1}:=-\sum_{k=2}^{p+1}\frac{\alpha_{1k}}{\beta^p_{k-2}} \Big/ \sum_{k=2}^{p+1}\frac{\alpha_{2k}}{\beta^p_{k-2}};$$

3° $t:=t_o+\nu\Delta t;$

$a_{p+1}:=\tilde{a}_{p+1};$

$k:=2;$

4° for $l=1,2,3$ and $i=1(1)N$:

$z_{li}:=x_{li};$

$u_{li}:=v_{li};$

$\mu:=0;$

5° for $l=1,2,3$ and $i=1(1)N$:

$$y_{li}:=z_{li}+\frac{H}{\beta_{k-2}}\Big(u_{li}-\frac{HG}{2\beta_{k-2}}\sum_{\substack{j=1\\ j\neq i}}^{N} m_j\frac{z_{li}-z_{lj}}{r^3_{ij}}\Big);$$

$$w_{li}:=u_{li}-\frac{HG}{\beta_{k-2}}\sum_{\substack{j=1\\ j\neq i}}^{N} m_j\frac{z_{li}-z_{lj}+\frac{H}{2\beta_{k-2}}(u_{li}-u_{lj})}{\tilde{r}^3_{ij}},$$

where $r_{ij}=[\sum_{l=1}^{3}(z_{li}-z_{lj})^2]^{1/2}$,

$\tilde{r}_{ij}=\{\sum_{l=1}^{3}[z_{li}-z_{lj}+\frac{H}{2\beta_{k-2}}(u_{li}-u_{lj})]^2\}^{1/2}$;

6° $\mu:=\mu+1$;
if $\mu\leq\beta_{k-2}-1$ then:
for l=1,2,3 and i=1(1)N:
$z_{li}:=y_{li}$;
$u_{li}:=w_{li}$;
go to 5°;

7° $x_{(k)li}:=y_{li}$;
$v_{(k)li}:=w_{li}$;
k:=k+1;
if $k\leq p+1$ then go to 4°;
$\bar{a}_{p+1}:=\tilde{a}_{p+1}$;

8° $$\Phi:=\sum_{i=1}^{N} m_i\{\sum_{l=1}^{3}[\sum_{k=2}^{p+1}(\alpha_{1k}+\alpha_{2k}a_{p+1})v_{(k)li}]^2-G\sum_{\substack{j=1\\j\neq i}}^{N}\frac{m_j}{r_{ij}}\}-2E^o;$$

$$\partial\Phi:=\sum_{i=1}^{N} m_i\{2\sum_{l=1}^{3}[\sum_{k=2}^{p+1}(\alpha_{1k}+\alpha_{2k}a_{p+1})v_{(k)li}][\sum_{k=2}^{p+1}\alpha_{2k}v_{(k)li}]$$
$$+G\sum_{\substack{j=1\\j\neq i}}^{N}\frac{m_j}{r_{ij}^3}\sum_{l=1}^{3}[\sum_{k=2}^{p+1}(\alpha_{1k}+\alpha_{2k}a_{p+1})(x_{(k)li}-x_{(k)lj})]$$
$$\times[\sum_{k=2}^{p+1}\alpha_{2k}(x_{(k)li}-x_{(k)lj})]\},$$

where $r_{ij}=\{\sum_{l=1}^{3}[\sum_{k=2}^{p+1}(\alpha_{1k}+\alpha_{2k}a_{p+1})(x_{(k)li}-x_{(k)lj})]^2\}^{1/2}$;

$a_{p+1}:=a_{p+1}-\Phi/\partial\Phi$;
if $|a_{p+1}-\bar{a}_{p+1}|\geq\varepsilon$ then:
$\bar{a}_{p+1}:=a_{p+1}$;
go to 8°;

9° for k=2(1)p:
$a_k:=\alpha_{1k}+\alpha_{2k}a_{p+1}$;
for l=1,2,3 and i=1(1)N:
$x_{li}^{\nu+1}:=\sum_{k=2}^{p+1}a_k x_{(k)li}$;
$v_{li}^{\nu+1}:=\sum_{k=2}^{p+1}a_k v_{(k)li}$;

printout $x_{li}^{\nu+1}$ and $v_{li}^{\nu+1}$ (l=1,2,3; i=1(1)N);
$\nu:=\nu+1$;
if $\nu \leq n-1$ then go to 3°;
stop.

Algorithm 12. The Runge-Kutta method of fourth order with energy conserving modification of polynomial extrapolation

1° compute H, E^o and ν as in Algorithm 11, 1°;
for k=4(1)p+1:
compute β_{k-4} from (1.3.51);

2° for k=4(1)p+1:

$$B_k := \prod_{\substack{l=4 \\ l \neq k}}^{p+1} \beta_{l-4};$$

for l=4(1)p-1 and s=p,p+1:

$$f_{ls} := \frac{B_s}{B_l} \prod_{\substack{i=4 \\ i \neq l}}^{p-1} \frac{B_s - B_i}{B_l - B_i}$$

$$A := 1 - \sum_{l=4}^{p-1} f_{l,p+1}$$

$$C := 1 - \sum_{l=4}^{p-1} f_{lp}$$

for l=4(1)p-1:
$\alpha_{1l} := -f_{lp}/C$;
$\alpha_{2l} := Af_{lp}/C - f_{l,p+1}$;
compute α_{1p}, α_{2p}, $\alpha_{1,p+1}$ and $\alpha_{2,p+1}$ as in Algorithm 11, 2°;

$$\tilde{a}_{p+1} := a_{p+1} := -\sum_{k=4}^{p+1} \frac{\alpha_{1k}}{\beta_{k-4}^p} \bigg/ \sum_{k=4}^{p+1} \frac{\alpha_{2k}}{\beta_{k-4}^p};$$

3° $t := t_o + \nu\Delta t$;
$a_{p+1} := \tilde{a}_{p+1}$;
k:=4;

4°, 7° as in Algorithm 11;

5° for l=1,2,3 and i=1(1)N:

$$A_{li}^{(1)} := -G \sum_{\substack{j=1 \\ j \neq i}}^{N} m_j \frac{z_{li} - z_{lj}}{r_{ij}^3}, \text{ where } r_{ij} = \Big[\sum_{l=1}^{3} (z_{li} - z_{lj})^2\Big]^{1/2};$$

$$A_{li}^{(2)} := -G \sum_{\substack{j=1 \\ j \neq i}}^{N} m_j \frac{z_{li} - z_{lj} + \frac{H}{2\beta_{k-4}}(u_{li} - u_{lj})}{\bar{r}_{ij}^3},$$

where $\bar{r}_{ij}=\{\sum_{l=1}^{3}[z_{li}-z_{lj}+\frac{H}{2\beta_{k-4}}(u_{li}-u_{lj})]^2\}^{1/2}$;

for l=1,2,3 and i=1(1)N:

$$A^{(3)}:=-G\sum_{\substack{j=1\\ j\neq i}}^{N} m_j \frac{z_{li}-z_{lj}+\frac{H}{2\beta_{k-4}}(u_{li}-u_{lj})+\frac{H^2}{4\beta_{k-4}^2}(A_{li}^{(1)}-A_{lj}^{(1)})}{\tilde{r}_{ij}^3},$$

where $\tilde{r}_{ij}=\{\sum_{l=1}^{3}[z_{li}-z_{lj}+\frac{H}{2\beta_{k-4}}(u_{li}-u_{lj})$

$+\frac{H^2}{4\beta_{k-4}^2}(A_{li}^{(1)}-A_{lj}^{(1)})]^2\}^{1/2}$;

$$A^{(4)}:=-G\sum_{\substack{j=1\\ j\neq i}}^{N} m_j \frac{z_{li}-z_{lj}+\frac{H}{\beta_{k-4}}(u_{li}-u_{lj})+\frac{H^2}{2\beta_{k-4}^2}(A_{li}^{2}-A_{lj}^{2})}{\hat{r}_{ij}^3},$$

where $\hat{r}_{ij}=\{\sum_{l=1}^{3}[z_{li}-z_{lj}+\frac{H}{\beta_{k-4}}(u_{li}-u_{lj})$

$+\frac{H^2}{2\beta_{k-4}^2}(A_{li}^{(2)}-A_{lj}^{(2)})]^2\}^{1/2}$;

$y_{li}:=z_{li}+\frac{H}{\beta_{k-4}}u_{li}+\frac{H}{6\beta_{k-4}}(A_{li}^{(1)}+A_{li}^{(2)}+A^{(3)})$;

$w_{li}:=u_{li}+\frac{H}{6\beta_{k-4}}(A_{li}^{(1)}+2A_{li}^{(2)}+2A^{(3)}+A^{(4)})$;

6° $\mu:=\mu+1$;
   if $\mu\leq\beta_{k-4}-1$ then:
      for l=1,2,3 and i=1(1)N:
         $z_{li}:=y_{li}$;
         $u_{li}:=w_{li}$;
      go to 5°;

8° compute Φ and $\partial\Phi$ as in Algorithm 11, 8° but begin the sumation from k=4;
   continue as in Algorithm 11, 8°;

9° for k=4(1)p:
      $a_k:=\alpha_{1k}+\alpha_{2k}a_{p+1}$;
   for l=1,2,3 and i=1(1)N:
      $x_{li}^{\nu+1}:=\sum_{k=4}^{p+1}a_k x_{(k)li}$;
      $v_{li}^{\nu+1}:=\sum_{k=4}^{p+1}a_k v_{(k)li}$.

Algorithm 13. The Gragg method with energy conserving modification of polynomial extrapolation

1° compute H, E^o and ν as in Algorithm 11, 1°;
for $k=1(1)p+1$:
compute β_{k-1} from (1.3.51);

2° for $k=1(1)p+1$:

$$B_k := \prod_{\substack{l=1 \\ l\neq k}}^{p+1} \beta_{l-1}^2;$$

for $l=1(1)p-1$ and $s=p,p+1$:
compute f_{ls} as in Algorithm 11, 2°;

compute A, C, α_{1l} and α_{2l} $(l=1(1)p+1)$ as in Algorithm 11, 2°;

$$\tilde{a}_{p+1} := a_{p+1} := -\sum_{k=1}^{p+1} \frac{\alpha_{1k}}{\beta_{k-1}^{2p}} \Big/ \sum_{k=1}^{p+1} \frac{\alpha_{2k}}{\beta_{k-1}^{2p}} ;$$

3° $t:=t_o+\nu\Delta t$;
$a_{p+1}:=\tilde{a}_{p+1}$;
$k:=1$;

4° for $l=1,2,3$ and $i=1(1)N$:

$z_{li}:=x_{li}$;

$w_{li}:=v_{li}$;

5° for $l=1,2,3$ and $i=1(1)N$:

$$y_{li}:=z_{li}+\frac{H}{2\beta_{k-1}}u_{li};$$

$$w_{li}:=u_{li}-\frac{HG}{2\beta_{k-1}}\sum_{\substack{j=1 \\ j\neq i}}^{N} m_j \frac{z_{li}-z_{lj}}{r_{ij}^3},$$

where $r_{ij}=[\sum_{l=1}^{3}(z_{li}-z_{lj})^2]^{1/2}$;

for $s=2(1)2\beta_{k-1}-1$:
for $l=1,2,3$ and $i=1(1)N$:
$\bar{x}_{li}:=y_{li}$;
$\bar{v}_{li}:=w_{li}$;
for $l=1,2,3$ and $i=1(1)N$:

$$y_{li}:=z_{li}+\frac{H}{\beta_{k-1}}\bar{v}_{li};$$

$$w_{li}:=u_{li}-\frac{HG}{\beta_{k-1}}\sum_{\substack{j=1 \\ j\neq i}}^{N} m_j \frac{\bar{x}_{li}-\bar{x}_{lj}}{\tilde{r}_{ij}^3},$$

where $\tilde{r}_{ij}=[\sum_{l=1}^{3}(\bar{x}_{li}-\bar{x}_{lj})^2]^{1/2}$;

for l=1,2,3 and i=1(1)N:

$z_{li}:=\bar{x}_{li}$;

$u_{li}:=\bar{v}_{li}$;

6° for l=1,2,3 and i=1(1)N:

$$x_{(k)li}:=\frac{1}{2}(y_{li}+z_{li})+\frac{H}{2\beta_{k-1}}(w_{li}+\frac{u_{li}}{2}-\frac{HG}{2\beta_{k-1}}\sum_{\substack{j=1\\j\neq i}}^{N}m_j\frac{y_{li}-y_{lj}}{r_{ij}^3};$$

$$v_{(k)li}:=\frac{1}{2}(w_{li}+u_{li})-\frac{HG}{2\beta_{k-1}}\sum_{\substack{j=1\\j\neq i}}^{N}m_j[\frac{y_{li}-y_{lj}}{r_{ij}^3}+\frac{z_{li}-z_{lj}+\frac{H}{\beta_{k-1}}(w_{li}-w_{lj})}{2\tilde{r}_{ij}^3}],$$

where $r_{ij}=[\sum_{l=1}^{3}(y_{li}-y_{lj})^2]^{1/2}$,

$\tilde{r}_{ij}=\{\sum_{l=1}^{3}[z_{li}-z_{lj}+\frac{H}{\beta_{k-1}}(w_{li}-w_{lj})]^2\}^{1/2}$;

7° k:=k+1;
if $k\leq p+1$ then go to 4°;
$a_{p+1}:=\tilde{a}_{p+1}$;

8° compute Φ and $\partial\Phi$ as in Algorithm 11, 8° but begin the sumation from k=1;
continue as in Algorithm 11, 8°;

9° for k=1(1)p:

$a_k:=\alpha_{1k}+\alpha_{2k}a_{p+1}$;

for l=1,2,3 and i=1(1)N:

$x_{li}^{\nu+1}:=\sum_{k=1}^{p+1}a_k x_{(k)li}$;

$v_{li}^{\nu+1}:=\sum_{k=1}^{p+1}a_k v_{(k)li}$;

printout $x_{li}^{\nu+1}$ and $v_{li}^{\nu+1}$ (l=1,2,3; i=1(1)N);
$\nu:=\nu+1$;
if $\nu\leq n-1$ then go to 3°;
stop.

The algorithms mentioned above applied to the problem of motion of two bodies (see Table V) yield the results presented in Tables XX, XXII and XXIV. Moreover, in Tables XXI, XXIII and XXV the results obtained from conventional polynomial extrapolation, i.e. without energy conserving modification, are given. The usefulness of our modification is self-evident.

Table XX. Solution of the two-body problem by the modified Euler method with energy conserving modification of polynomial extrapolation (of order 4; $\Delta t=0.01$, $\varepsilon=0.000001$)

ν	i	l	x_{li}	v_{li}	characteristics (*) constants of motion
20	1	1	0.309019065309	-5.975646234908	α=20, β=10
		2	0.951057450006	1.941624250917	Δ=0.000003%
	2	1	0.000002075280	0.000017947147	$c_1=2.9\times10^{-14}$
		2	0.000000917772	0.000013039365	c_2=6.283185307180
					c_4=1.000000000000
					c_5=0
					c_9=6.283185307180
					c_{10}=-19.73909255381
60	1	1	-0.809011753515	3.693151144902	α=20, β=10
		2	-0.587772022775	-5.083170028295	Δ=0.000023%
	2	1	0.000005433153	-0.000011091943	$c_1=4.8\times10^{-14}$
		2	0.000013087786	0.000034137505	$c_5=-7.0\times10^{-14}$
					c_2, c_4, c_9, c_{10} as above
100	1	1	1.000000000000	0.000002383819	α=20, β=10
		2	0.000018491351	6.283185307180	Δ=0.000038%
	2	1	0.000000000000	-0.000000000007	$c_1=3.5\times10^{-14}$
		2	0.000018870749	0.000000000000	c_2=6.283185307181
					$c_5=-1.92\times10^{-13}$
					c_4, c_9, c_{10} as above

CPU time 1.44 sec

(*) Explanation of notations - see p.69

Table XXI. Solution of the two-body problem by the modified Euler method with conventional polynomial extrapolation (order, Δt and ε as in Table XX)

ν	i	l	x_{li}	v_{li}	characteristics constants of motion
20	1	1	0.308983421080	-5.976466277093	α=12, β=6
		2	0.950971031317	1.940995341376	Δ=0.013992%
	2	1	0.000002075387	0.000017949610	c_1=1.7 10
		2	0.000000918032	0.000013041254	c_2=6.283185307180
					c_4=1.000000000000
					$c_5=0.2\times10^{-14}$
					c_9=6.283185205176

Table XXI. (cont.)

ν	i	l	x_{li}	v_{li}	characteristics constants of motion
					c_{10}=-19.73909269874
60	1	1	-0.808255293962	3.698481471180	α=12, β=6
		2	-0.588397012025	-5.081190948985	Δ=0.091070%
	2	1	0.000005430881	-0.000011107952	c_1=-2.5×10^{-14}
		2	0.000013089664	0.000034131561	c_2=6.283185307180
					c_4=1.000000000000
					c_5=3.0×10^{-14}
					c_9=6.283185061272
					c_{10}=-19.73909279943
100	1	1	0.999999218296	-0.008290179005	α=12, β=6
		2	0.001338421495	6.283179558910	Δ=0.136843%
	2	1	0.000000000002	0.000000024899	c_1=-0.3×10^{-14}
		2	0.000018866785	0.000000000017	c_2=6.283185307180
					c_4=1.000000000000
					c_5=-4.7×10^{-14}
					c_9=6.283185586669
					c_{10}=-19.73909079773

CPU time 1.07 sec

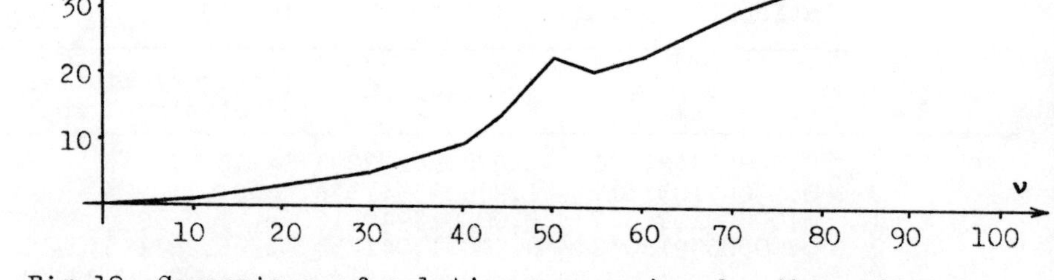

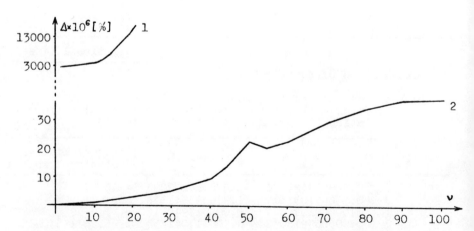

Fig.18. Comparison of solution accuracies for the modified Euler method with conventional (1) and energy conserving (2) extrapolation (accuracies of order 4)

Table XXII. Solution of the two-body problem by the fourth order method of Runge-Kutta with energy conserving modification of polynomial extrapolation (solution of order 7; $\Delta t=0.01$, $\varepsilon=0.000001$)

ν	i	l	x_{li}	v_{li}	characteristics constants of motion
20	1	1	0.309019023071	-5.975646798815	$\alpha=64$, =16
		2	0.951057407055	1.941623611357	$\Delta=0.000008\%$
	2	1	0.000002075280	0.000017947149	$c_1=1.27\times10^{-13}$
		2	0.000000917772	0.000013039367	$c_2=6.283185307180$
					$c_4=1.000000000000$
					$c_5=2.1\times10^{-14}$
					$c_9=6.283185307180$
					$c_{10}=-19.73909255381$
60	1	1	-0.809011178246	3.693155361041	$\alpha=64$, $\beta=16$
		2	-0.587772513081	5.083168341365	$\Delta=0.000049\%$
	2	1	0.000005433151	-0.000011091956	$c_2=6.283185307179$
		2	0.000013087788	0.000034137500	c_4, c_9, c_{10} as previously
100	1	1	1.000000000000	-0.000004596293	$\alpha=64$, $\beta=16$
		2	0.000019602270	6.283185307177	$\Delta=0.000073\%$
	2	1	0.000000000000	0.000000000014	c_2, c_4, c_9, c_{10}
		2	0.000018870746	0.000000000000	as above
					CPU time 5.43 sec

Table XXIII. Solution of the two-body problem by the fourth order Runge-Kutta method with conventional polynomial extrapolation (order, Δt and ε as in Table XXII)

ν	i	l	x_{li}	v_{li}	characteristics constants of motion
20	1	1	0.309019021580	-5.975646849668	$\alpha=40$, $\beta=10$
		2	0.951057401036	1.941623575210	$\Delta=0.000010\%$
	2	1	0.000002075280	0.000017947149	$c_1=-4.1\times10^{-14}$
		2	0.000000917772	0.000013039368	$c_2=6.283185307180$
					$c_4=1.000000000000$
					$c_5=-0.3\times10^{-14}$
					$c_9=6.283185305516$
					$c_{10}=-19.73909256427$
60	1	1	-0.809011128258	3.693155710199	$\alpha=40$, $\beta=10$
		2	-0.587772552341	-5.083168216378	$\Delta=0.000055\%$
	2	1	0.000005433151	-0.000011091957	$c_1=-1.60\times10^{-13}$
					$c_2=6.283185307180$
					$c_4=1.000000000000$
					$c_9=-4.6\times10^{-14}$

Table XXIII. cont.

ν	i	l	x_{li}	v_{li}	characteristics constants of motion
					c_9=6.283185302188
					c_{10}=-19.73909258518
100	1	1	0.999999997351	-0.000005173828	α=40, β=10
		2	0.000019694188	6.283185315497	Δ=0.000083%
	2	1	0.000000000000	0.000000000016	$c_1=-0.2\times10^{-14}$
		2	0.000018870745	0.000000000000	c_2=6.283185307180
					c_4=1.000000000000
					$c_5=-3.61\times10^{-13}$
					c_9=6.283185298860
					c_{10}=-19.73909260609
					CPU time 2.57 sec

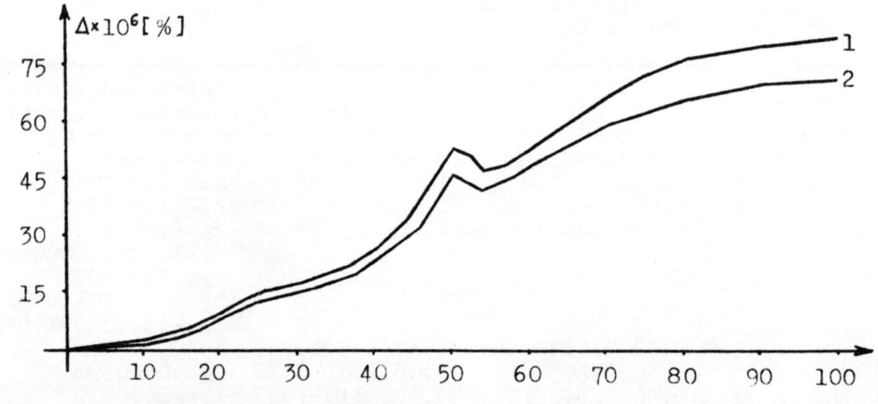

Fig.19. Comparison of solution accuracies for the Runge-Kutta method with conventional (1) and energy conserving extrapolation (2)

Table XXIV. Solution of the two-body problem by the Gragg method with energy conserving modification of polynomial extrapolation (solution of order 8; $\Delta t=0.1$, $\varepsilon=0.000001$)

ν	i	l	x_{li}	v_{li}	characteristics constants of motion
2	1	1	0.309018999306	-5.975646852646	$\alpha=42$, $\beta=16$
		2	0.951057409959	1.941623538830	$\Delta=0.000010\%$
	2	1	0.000002075280	0.000017947149	$c_1=-0.3\times10^{-14}$
		2	0.000000917772	0.000013039368	$c_2=6.283185307180$
					$c_4=1.000000000000$
					$c_5=0$
					$c_9=6.283185307180$
					$c_{10}=-19.73909255381$
6	1	1	-0.809011167978	3.693155431533	$\alpha=42$, $\beta=16$
		2	-0.587772549166	-5.083168189937	$\Delta=0.000052\%$
	2	1	0.000005433151	-0.000011091956	$c_1=-1.0$ 10
		2	0.000013087788	0.000034137499	$c_5=0.8$ 10
					c_2, c_4, c_9, c_{10} as above
10	1	1	1.000000000000	-0.000004714171	$\alpha=42$, $\beta=16$
		2	0.000019621031	6.283185307178	$\Delta=0.000075\%$
	2	1	0.000000000000	0.000000000014	$c_1=0.1$ 10
		2	0.000018870746	0.000000000000	$c_5=-1.0$ 10
					c_2, c_4, c_9, c_{10} as above
					CPU time 0.26 sec

Table XXV. Solution of the two-body problem by the Gragg method with conventional polynomial extrapolation (order, Δt and ε as in Table XXIV)

ν	i	l	x_{li}	v_{li}	characteristics constants of motion
2	1	1	0.309019591761	-5.975645237232	$\alpha=28$, $\beta=10$
		2	0.951057379370	1.941626243081	$\Delta=0.000044\%$
	2	1	0.000002075278	0.000017947144	$c_1=0.8\times10^{-14}$
		2	0.000000917772	0.000013039360	$c_2=6.283185307180$
					$c_4=1.000000000000$
					$c_5=0$
					$c_9=6.283185574018$
					$c_{10}=-19.73909087721$
6	1	1	-0.809012467532	3.693145945476	$\alpha=28$, $\beta=10$
		2	-0.587771032469	-5.083174829822	$\Delta=0.000141\%$
	2	1	0.000005433155	-0.000011091927	$c_1=0.9\times10^{-14}$
		2	0.000013087783	0.000034137519	$c_2=6.283185307180$

Table XXV. (cont.)

ν	i	l	x_{li}	v_{li}	characteristics constants of motion
					c_4 =1.000000000000
					c_5 =-0.2×10^{-14}
					c_9 =6.283186107695
					c_{10}=-19.73908752403
10	1	1	1.000000424681	0.000018057702	α=28, β=10
		2	0.000015996774	6.283183972969	Δ=0.000319%
	2	1	-0.000000000001	-0.000000000054	c_1 =0.5×10^{-14}
		2	0.000018870756	0.000000000004	c_2 =6.283185307180
					c_4 =1.000000000000
					c_5 =-0.1×10^{-14}
					c_9 =6.283186641370
					c_{10}=-19.73908417085
					CPU time 0.17 sec

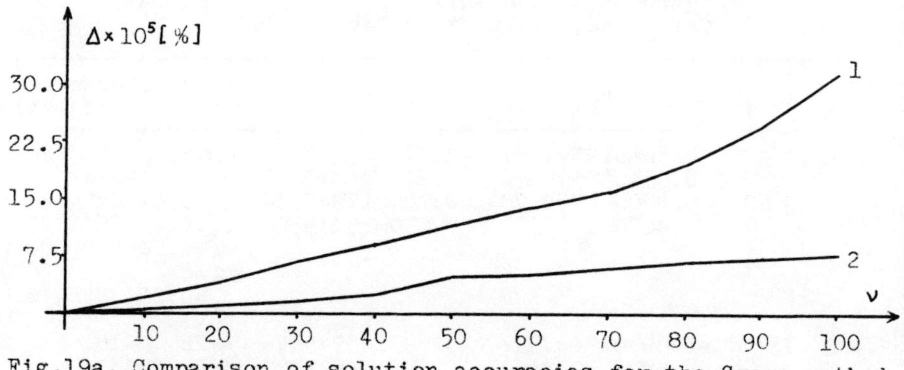

Fig.19a. Comparison of solution accuracies for the Gragg method with conventional (1) and energy conserving (2) extrapolation

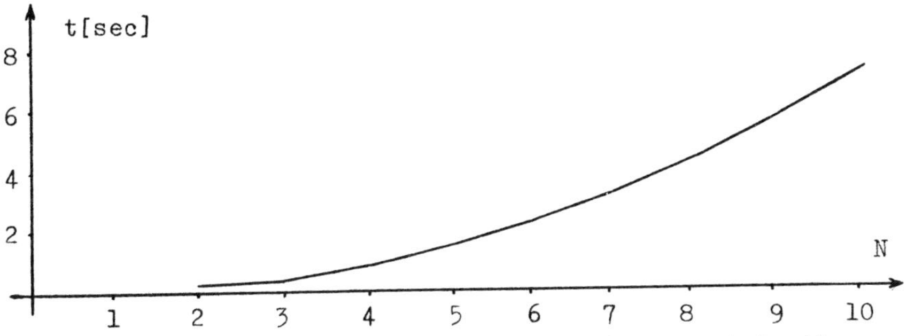

Fig.20. Increase of computational time with respect to the number of bodies for Gragg's method with energy conserving modification of polynomial extrapolation (Δt=0.1, ε=0.000001)

Table XXVI. Solution of the six-body problem by the Algorithm 8 (solution of order 8; Δt=0.1, ε=0.000001, the data at the initial moment as in Table XIV)

ν	i	l	x_{li}	v_{li}	characteristics coefficients constants of motion
4	1	1	0.797541084152	-3.846015196691	α=42, β=16
		2	0.591563299985	5.025323853303	a_1 = 0.0002027
		3	-0.000000976194	-0.000006751932	a_2 =-0.0617092
	2	1	0.316844297249	-2.785262940786	a_3 = 1.0176904
		2	5.106515144430	0.300169990818	a_4 =-2.8531934
		3	-0.072775915477	-0.049051702077	a_5 = 2.8970094
	3	1	-7.365338029930	-1.332779550353	c_1 = -954.554477572
		2	5.508618667294	-1.636083804869	c_2 = 117.700248203
		3	0.198595669750	0.081406895179	c_3 = -10.429703622
	4	1	-13.812278783563	0.948344507109	c_4 = -541.574672823
		2	-12.453555964992	-1.133405423328	c_5 = 1451.693675512
		3	0.133076196103	-0.016496547734	c_6 = 14.887435923
	5	1	-7.516402712151	1.100593438472	c_7 = 19.833730446
		2	-29.329386477820	-0.277705512848	c_8 = 111.933710616
		3	0.776226732621	-0.019775209985	c_9 = 7394.867175401
	6	1	0.000019665856	0.000085974731	c_{10}=-1447.506534332
		2	0.000113764559	0.000583392275	
		3	-0.000001136890	-0.000006023630	
7	1	1	-0.834716656412	-3.464366346937	α=42, β=16
		2	0.528884558636	-5.331126875145	a_1 = 0.0002614
		3	-0.000004456721	-0.000014717421	a_2 =-0.0699379

Table XXVI. (cont.)

ν	i	l	x_{li}	v_{li}	characteristics coefficients constants of motion
	2	1	-0.519257080220	-2.776594114457	a_3 = 1.0891035
		2	5.129109235998	-0.148168784024	a_4 =-2.9735671
		3	-0.086471074620	-0.042076110339	a_5 = 2.9541399
	3	1	-7.748088899089	-1.218092491604	c_1 = -954.554477572
		2	5.005636519252	-1.715739678386	c_2 = 117.700248203
		3	0.222546466195	0.078208937528	c_3 = -10.429703622
	4	1	-13.523988098595	0.973493704642	c_4 = -541.574672823
		2	-12.790106582336	-1.110172262470	c_5 = 1451.693675512
		3	0.128090954515	-0.016736761872	c_6 = 14.887435923
	5	1	-7.185750880984	1.103728611824	c_7 = 19.833730446
		2	-29.410818907330	-0.265171261328	c_8 = 111.933710616
		3	0.770244611979	-0.020105176924	c_9 = 7394.867175398
	6	1	0.000042321224	0.000042511390	c_{10}=-1447.506534332
		2	0.000361118320	0.001063582781	
		3	-0.000003779140	-0.000011740807	
10	1	1	-0.259016293192	5.973304187471	α=42, β=16
		2	-0.980382146510	-1.625497541434	a_1 = 0.0001799
		3	-0.000008452967	-0.000012215906	a_2 =-0.0585237
	2	1	-1.342026032236	-2.697120365949	a_3 = 0.9900452
		2	5.019013993143	-0.582554112169	a_4 =-2.8065946
		3	-0.097929676839	-0.034181641069	a_5 = 2.8748933
	3	1	-8.095746401646	-1.098934310873	c_1 = -954.554477572
		2	4.480021356115	-1.786932576785	c_2 = 117.700248203
		3	0.245488434425	0.074685035122	c_3 = -10.429703622
	4	1	-13.228242976545	0.998038763826	c_4 = -541.574672823
		2	-13.119604311790	-1.086389001348	c_5 = 1451.693675512
		3	0.123035124032	-0.016967125784	c_6 = 14.887435923
	5	1	-6.854179662278	1.106722613447	c_7 = 19.833730446
		2	-29.488485660737	-0.252601270895	c_8 = 111.933710616
		3	0.764163879045	-0.020432618106	c_9 = 7394.867175398
	6	1	0.000035214222	-0.000096984950	c_{10}=-1447.506534332
		2	0.000745447444	0.001485774741	
		3	-0.000008259849	-0.000018242567	
					CPU time 2.19 sec

Let us note that in the last example the constants of motion are numerically stable and that the Gragg method with the modification of polynomial extrapolation presented in this section is not too consumptive of computer time.

In this chapter we have presented a lot of numerical methods for solving the general N-body problem. It is very difficult to point at the best method, because it depends on the point of view. If we omit the problem of conservation of the constants of motion and we want to find the solution with very high accuracy, then the rational extrapolation method of Gragg-Bulirsch-Stoer seems to be the best, especially when $N \gg 2$. The methods conserving constants of motion yield a better approximation of the exact solution in comparison with adequate conventional methods, but they are more expensive taking into account the time of computations. The author prefers the application of these methods since they have the properties of continuous mechanics solutions.

Chapter 3

The Relative Motion of N Bodies

3.1. EQUATIONS OF MOTION

In the previous chapter we have considered the equations of absolute motion of N material points, i.e. the equations of motion in some inertial frame of reference. But the basic problem in celestial mechanics is the study of mutual positions of celestial bodies. Therefore, we usually consider a relative motion of those bodies, i.e. the motion with respect to some central body of the system. Usually it is a body with the greatest mass. For instance, in the solar system we determine the motion of planets with respect to the sun.

If we put the origin of the rectangular coordinate system at the given material point, then such a frame will not be an inertial one, because, as we have shown in Sect.2.1, not the central body, but the center of mass moves in a straight line with uniform velocity, i.e. inertially. A consideration of motion of material points in an uninertial frame of reference implies that the constants of motion (2.1.6), (2.1.7),(2.1.12) and (2.1.16) are not conserved. Nevertheless, taking into account the great weight of this problem in practice, all this chapter is devoted to the study of relative motion of bodies.

Let us consider a system of N material points with the masses m_i (i=1(1)N). Let us put the origin of the frame of reference at the material point with the mass m_N, and let the axes ξ_1, ξ_2 and ξ_3 of this frame be parallel to the axes x_1, x_2 and x_3 of some inertial frame. The rectangular coordinates ξ_{li} (l=1,2,3; i=1(1)N-1) of material points with masses m_i are given by

$$\xi_{li} = x_{li} - x_{lN}; \quad l=1,2,3; \; i=1(1)N-1. \tag{3.1.1}$$

From (2.1.1)-(2.1.2) we have

$$m_i\dot{v}_{li} = -Gm_i \sum_{\substack{j=1\\ j\neq i}}^{N-1} m_j \frac{x_{li}-x_{lj}}{r_{ij}^3} - Gm_i m_N \frac{x_{li}-x_{lN}}{r_{iN}^3}.$$

Taking into account (3.1.1), we get

$$\dot{\eta}_{li}+\dot{v}_{lN} = -G \sum_{\substack{j=1\\ j\neq i}}^{N-1} m_j \frac{\xi_{li}-\xi_{lj}}{r_{ij}^3} - Gm_i m_N \frac{\xi_{li}}{r_{iN}^3}, \tag{3.1.2}$$

where

$$r_{ij}=[\sum_{l=1}^{3}(\xi_{li}-\xi_{lj})^2]^{1/2}, \quad r_{iN}=(\sum_{l=1}^{3}\xi_{li}^2)^{1/2}, \; \eta_{li}=\dot{\xi}_{li}.$$

Moreover, from the equations (2.1.1)-(2.1.2) and (3.1.1) we obtain

$$\dot{v}_{lN} = G \sum_{j=1}^{N-1} m_j \frac{\xi_{lj}}{r_{jN}^3}. \tag{3.1.3}$$

Substitution of (3.1.3) in (3.1.2) and the relation (2.1.1) yield the equations of relative motion as follows

$$\begin{aligned} &\dot{\xi}_{li} = \eta_{li}, \quad \xi_{li}(t_o) = \xi_{li}^o, \\ &\dot{\eta}_{li} = -G[(m_N+m_i)\frac{\xi_{li}}{r_{iN}^3} + \sum_{\substack{j=1\\ j\neq i}}^{N-1} m_j(\frac{\xi_{li}-\xi_{lj}}{r_{ij}^3} + \frac{\xi_{lj}}{r_{jN}^3})], \\ &\eta_{li}(t_o) = \eta_{li}^o; \qquad l=1,2,3; \; i=1(1)N-1. \end{aligned} \tag{3.1.4}$$

The system (3.1.4) of differential equations is of order 6N-6. To solve this system for N=2 we have to find six constants of motion. These constants can be determined from initial conditions, i.e. from coordinates and components of velocity at the initial moment. That means we can get an exact analytical solution of the relative two-body problem.

Let us find the analytical solution of the equations of relative motion for two bodies P_1 and P_2. Based on this solution, we will verify numerical methods applied for solving the equations of relative motion in the next sections. Let us take the frame of reference in such a way that its origin coincides with the material point P_2. From (3.1.4) we get the equations of motion of the material point P_1 as follows (the axes of the frame are denoted further by x_1, x_2 and x_3)

$$\dot{x}_l = v_l, \quad x_l(t_o) = x_l^o,$$
$$\dot{v}_l = -G(m_1+m_2)\frac{x_l}{r^3}, \quad v_l(t_o) = v_l^o; \quad l=1,2,3; \tag{3.1.5}$$

where

$$r = (\sum_{l=1}^{3} x_l^2)^{1/2}.$$

At first, let us note that

$$x_p\dot{v}_q - x_q\dot{v}_p = 0,$$

where $(p,q)\epsilon\{(2,3),(3,1),(1,2)\}$. Hence, after integration, we obtain

$$x_p v_q - x_q v_p = d_l, \tag{3.1.6}$$

where d_l=const and $l=1$ for $(p,q)=(2,3)$, $l=2$ for $(p,q)=(3,1)$ and $l=3$ for $(p,q)=(1,2)$. If we multiply the equations (3.1.6) by x_1, x_2 and x_3 in turn and add them together, we get

$$\sum_{l=1}^{3} d_l x_l = 0. \tag{3.1.7}$$

The equation (3.1.7) is an equation of plane going over the origin of the coordinate system. Taking

$$\kappa_1 = d_1/d_3, \ \kappa_2 = d_2/d_3,$$

we have

$$\kappa_1 x_1 + \kappa_2 x_2 + x_3 = 0.$$

The last equation with two constants κ_1 and κ_2 determine the plane of motion of the material point P_1.

If in the second equation of (3.1.5) we multiply by v_1, v_2 and v_3 in turn and then we add these equations together, we obtain

$$\sum_{l=1}^{3} v_l\dot{v}_l = -\frac{G(m_1+m_2)}{r^3}\sum_{l=1}^{3} x_l v_l.$$

We can rewrite this equation in the form

$$\frac{1}{2}\frac{d}{dt}\sum_{l=1}^{3} v_l^2 = G(m_1+m_2)\frac{d}{dt}(\frac{1}{r}).$$

Integration yields

$$\frac{1}{2}\sum_{l=1}^{3} v_l^2 - \frac{G(m_1+m_2)}{r} = d_4, \tag{3.1.8}$$

where d_4=const. Thus, we obtain the fourth constant of motion. The remaining two constants we get if we consider the motion in the orbit plane using angular coordinates (see e.g. [94, p.135]).

From the analytical theory it is known ([22],[72],[83],[94]) that the motion of P_1 with respect to P_2 goes on a circle if at the initial moment the following conditions are fulfilled

$$\sum_{l=1}^{3} x_l^o v_l^o = 0,$$

$$\sum_{l=1}^{3} (v_l^o)^2 = \frac{G(m_1+m_2)}{r^o},$$

where

$$r^o = [\sum_{l=1}^{3} (x_l^o)^2]^{1/2}.$$

So, let us assume that the material points P_1 and P_2 with masses $m_1=1$ and $m_2=332958$ respectively fulfil at the initial moment $t_o=0$ the conditions: $x_1^o=1$, $v_2^o=\sqrt{G(m_1+m_2)}=6.283185491803 \simeq 2\pi$; $x_l^o, v_l^o=0$ for other l, where the gravitational constant G has the value (2.1.4). It is easy to check that the solution of this problem has the form

$$x_1 = \cos Mt, \quad x_2 = \sin Mt, \quad x_3 = 0,$$
$$v_1 = -M\sin Mt, \quad v_2 = M\cos Mt, \quad v_3 = 0, \tag{3.1.9}$$

where $M=\sqrt{G(m_1+m_2)}$. In Table XXVII the values of solution at the moments $t_\nu=\nu/100$ for ν=20(20)100 are given. We will apply this example to verify numerical methods for solving the problem of the relative motion of N bodies.

Table XXVII. Exact solution of the relative motion problem of two bodies

ν	l	x_l	v_l	constants of motion
20	1	0.309016959257	-5.975664576764	$d_3=\sqrt{G(m_1+m_2)}$
	2	0.951056527706	1.941610875128	=6.283185491803
40	1	-0.809017037783	-3.693163394108	$d_4=-G(m_1+m_2)/2$
	2	0.587785192547	-5.083204114417	=-19.73920996220
60	1	-0.809016929264	3.693164332587	
	2	-0.587785341911	-5.083203432572	
80	1	0.309017134845	5.975664218297	
	2	-0.951056470654	1.941611978375	
100	1	1.000000000000	-0.000001160023	
	2	0.000000184623	6.283185491803	

3.2. THE APPLICATION OF CONVENTIONAL NUMERICAL METHODS

Algorithms of conventional numerical methods applied to the problem of relative motion of N bodies differ insignificantly from those given in Sect.2.2. Only the equations of motion, i.e. differential equations, are different. Therefore, we restrict ourselves to the presentation of four algorithms. At the same time we will refer to the algorithms given previously.

Let us assume that at the initial moment t_o and in a rectangular coordinate system with the origin in a material point having the mass m_N, the coordinates x^o_{li} and components of velocities v^o_{li} ($l=1,2,3$; $i=1(1)N-1$) of N-1 material points with masses m_i are given. Let the motion of these points be described by the system of differential equations (3.1.4). What we ought to do is find the solution of this system at the moments $t_\nu=t_o+\nu\Delta t$, where $\Delta t=(t_n-t_o)/n$ $(\nu=1(1)n)$.

Algorithm 14. The Runge-Kutta method of fourth order with an automatic step size correction (compare Algorithm 2)

$1^o,2^o,4^o,5^o,7^o,8^o$ - as in Algorithm 2, but the summation with respect to i ought to be done from 1 to N-1;

3° for $l=1,2,3$ and $i=1(1)N-1$:

$$A^{(1)}_{li}:=-G\{(m_N+m_i)\frac{x^k_{li}}{(r^k_{iN})^3}+\sum_{\substack{j=1\\j\neq i}}^{N-1} m_j[\frac{x^k_{li}-x^k_{lj}}{(r^k_{ij})^3}+\frac{x^k_{lj}}{(r^k_{jN})^3}]\},$$

$$\text{where } r^k_{iN}=[\sum_{l=1}^{3}(x^k_{li})^2]^{1/2},$$

$$r^k_{ij}=[\sum_{l=1}^{3}(x^k_{li}-x^k_{lj})^2]^{1/2};$$

$$A^{(2)}_{li}:=-G\{(m_N+m_i)\frac{x^k_{li}+\frac{H}{2}v^k_{li}}{(\bar{r}^k_{iN})^3}+\sum_{\substack{j=1\\j\neq i}}^{N-1} m_j[\frac{x^k_{li}-x^k_{lj}+\frac{H}{2}(v^k_{li}-v^k_{lj})}{(\bar{r}^k_{ij})^3}+\frac{x^k_{lj}+\frac{H}{2}v^k_{lj}}{(\bar{r}^k_{jN})^3}]\},$$

$$\text{where } \bar{r}^k_{iN}=[\sum_{l=1}^{3}(x^k_{li}+\frac{H}{2}v^k_{li})^2]^{1/2},$$

$$\text{where } \bar{r}^k_{ij}=\{\sum_{l=1}^{3}[x^k_{li}-x^k_{lj}+\frac{H}{2}(v^k_{li}-v^k_{lj})]^2\}^{1/2};$$

for l=1,2,3 and i=1(1)N-1:

$$A^{(3)}:=-G\{(m_N+m_i)\frac{x^k_{li}+\frac{H}{2}v^k_{li}+\frac{H^2}{4}A^{(1)}_{li}}{(\tilde{r}^k_{iN})^3}$$

$$+\sum_{\substack{j=1\\j\neq i}}^{N-1} m_j[\frac{x^k_{li}-x^k_{lj}+\frac{H}{2}(v^k_{li}-v^k_{lj})+\frac{H^2}{4}(A^{(1)}_{li}-A^{(1)}_{lj})}{(\tilde{r}^k_{ij})^3}$$

$$+\frac{x^k_{lj}+\frac{H}{2}v^k_{lj}+\frac{H^2}{4}A^{(1)}_{lj}}{(\tilde{r}^k_{jN})^3}]\},$$

$$\text{where } \tilde{r}^k_{iN}=[\sum_{l=1}^{3}(x^k_{li}+\frac{H}{2}v^k_{li}+\frac{H^2}{4}A^{(1)}_{li})^2]^{1/2},$$

$$\tilde{r}^k_{ij}=\{\sum_{l=1}^{3}[x^k_{li}-x^k_{lj}+\frac{H}{2}(v^k_{li}-v^k_{lj})+\frac{H^2}{4}(A^{(1)}_{li}-A^{(1)}_{lj})]^2\}^{1/}$$

$$A^{(4)}:=-G\{(m_N+m_i)\frac{x^k_{li}+Hv^k_{li}+\frac{H^2}{2}A^{(2)}_{li}}{(\hat{r}^k_{iN})^3}$$

$$+\sum_{\substack{j=1\\j\neq i}}^{N-1} m_j[\frac{x^k_{li}-x^k_{lj}+H(v^k_{li}-v^k_{lj})+\frac{H^2}{2}(A^{(2)}_{li}-A^{(2)}_{lj})}{(\hat{r}^k_{ij})^3}$$

$$+\frac{x^k_{lj}+Hv^k_{lj}+\frac{H^2}{2}A^{(2)}_{lj}}{(\hat{r}^k_{jN})^3}]\},$$

$$\text{where } \hat{r}^k_{iN}=[\sum_{l=1}^{3}(x^k_{li}+Hv^k_{li}+\frac{H^2}{2}A^{(2)}_{li})^2]^{1/2},$$

$$\hat{r}^k_{ij}=\{\sum_{l=1}^{3}[x^k_{li}-x^k_{lj}+H(v^k_{li}-v^k_{lj})+\frac{H^2}{2}(A^{(2)}_{li}-A^{(2)}_{lj})]^2\}^{1/2}$$

$$x^{k+1}_{li}:=x^k_{li}+H[v^k_{li}+\frac{H}{6}(A^{(1)}_{li}+A^{(2)}_{li}+A^{(3)})];$$

$$v^{k+1}_{li}:=v^k_{li}+\frac{H}{6}(A^{(1)}_{li}+2A^{(2)}_{li}+2A^{(3)}+A^{(4)});$$

6° for l=1,2,3 and i=1(1)N-1:

$$A^{(1)}_{li}:=-G[(m_N+m_i)\frac{y_{li}}{r^3_{iN}}+\sum_{\substack{j=1\\j\neq i}}^{N-1} m_j(\frac{y_{li}-y_{lj}}{r^3_{ij}}+\frac{y_{lj}}{r^3_{jN}})],$$

$$\text{where } r_{iN}=(\sum_{l=1}^{3}y^2_{li})^{1/2},$$

$$r_{ij}=[\sum_{l=1}^{3}(y_{li}-y_{lj})^2]^{1/2};$$

$$A_{li}^{(2)}:=-G\{(m_N+m_i)\frac{y_{li}+\frac{H}{4}w_{li}}{\bar{r}_{iN}^3}$$
$$+\sum_{\substack{j=1\\j\neq i}}^{N-1} m_j[\frac{y_{li}-y_{lj}+\frac{H}{4}(w_{li}-w_{lj})}{\bar{r}_{ij}^3}+\frac{y_{lj}+\frac{H}{4}w_{lj}}{\bar{r}_{jN}^3}]\},$$

where $\bar{r}_{iN}=[\sum_{l=1}^{3}(y_{li}+\frac{H}{4}w_{li})^2]^{1/2}$,

$$\bar{r}_{ij}=\{\sum_{l=1}^{3}[y_{li}-y_{lj}+\frac{H}{4}(w_{li}-w_{lj})]^2\}^{1/2};$$

for l=1,2,3 and i=1(1)N-1:

$$A^{(3)}:=-G\{(m_N+m_i)\frac{y_{li}+\frac{H}{4}w_{li}+\frac{H^2}{16}A_{li}^{(1)}}{\tilde{r}_{iN}^3}$$
$$+\sum_{\substack{j=1\\j\neq i}}^{N-1} m_j[\frac{y_{li}-y_{lj}+\frac{H}{4}(w_{li}-w_{lj})+\frac{H^2}{16}(A_{li}^{(1)}-A_{lj}^{(1)})}{\tilde{r}_{ij}^3}$$
$$+\frac{y_{lj}+\frac{H}{4}w_{lj}+\frac{H^2}{16}A_{lj}^{(1)}}{\tilde{r}_{jN}^3}]\},$$

where $\tilde{r}_{iN}=[\sum_{l=1}^{3}(y_{li}+\frac{H}{4}w_{li}+\frac{H^2}{16}A_{li}^{(1)})^2]^{1/2}$,

$$\tilde{r}_{ij}=\{\sum_{l=1}^{3}[y_{li}-y_{lj}+\frac{H}{4}(w_{li}-w_{lj})$$
$$+\frac{H^2}{16}(A_{li}^{(1)}-A_{lj}^{(1)})]^2\}^{1/2};$$

$$A^{(4)}:=-G\{(m_N+m_i)\frac{y_{li}+\frac{H}{2}w_{li}+\frac{H^2}{8}A_{li}^{(2)}}{\hat{r}_{iN}^3}$$
$$+\sum_{\substack{j=1\\j\neq i}}^{N-1} m_j[\frac{y_{li}-y_{lj}+\frac{H}{2}(w_{li}-w_{lj})+\frac{H^2}{8}(A_{li}^{(2)}-A_{lj}^{(2)})}{\hat{r}_{ij}^3}$$
$$+\frac{y_{lj}+\frac{H}{2}w_{lj}+\frac{H^2}{8}A_{lj}^{(2)}}{\hat{r}_{jN}^3}]\},$$

where $\hat{r}_{iN}=[\sum_{l=1}^{3}(y_{li}+\frac{H}{2}w_{li}+\frac{H^2}{8}A_{li}^{(2)})^2]^{1/2}$,

$$\hat{r}_{ij}=\{\sum_{l=1}^{3}[y_{li}-y_{lj}+\frac{H}{2}(w_{li}-w_{lj})+\frac{H^2}{8}(A_{li}^{(2)}-A_{lj}^{(2)})]^2\}^{1/2}.$$

Algorithm 15. Predictor-corrector method of seventh order (compare Algorithm 5)

$1^o,2^o,5^o,7^o$ - as in Algorithm 5, but the summation with respect to i ought to be done from 1 to N-1;

3° for l=1,2,3 and i=1(1)N-1:

$$x_{li}^{k+1}:=x_{li}^{k}+\frac{\Delta t}{1.440}(4.227v_{li}^{k}-7.673v_{li}^{k-1}+9.482v_{li}^{k-2}-6.798v_{li}^{k-3}+2.627v_{li}^{k-4}-0.425v_{li}^{k-5});$$

$$v_{li}^{k+1}:=v_{li}^{k}-\frac{\Delta t\times G}{1.440}((m_N+m_i)[4.227\frac{x_{li}^{k}}{(r_{iN}^{k})^3}-7.673\frac{x_{li}^{k-1}}{(r_{iN}^{k-1})^3}+9.482\frac{x_{li}^{k-2}}{(r_{iN}^{k-2})^3}-6.798\frac{x_{li}^{k-3}}{(r_{iN}^{k-3})^3}+2.627\frac{x_{li}^{k-4}}{(r_{iN}^{k-4})^3}-0.425\frac{x_{li}^{k-5}}{(r_{iN}^{k-5})^3}]$$

$$+\sum_{\substack{j=1\\j\neq i}}^{N-1}m_j\{4.227[\frac{x_{li}^{k}-x_{lj}^{k}}{(r_{ij}^{k})^3}+\frac{x_{lj}^{k}}{(r_{jN}^{k})^3}]-7.673[\frac{x_{li}^{k-1}-x_{lj}^{k-1}}{(r_{ij}^{k-1})^3}+\frac{x_{lj}^{k-1}}{(r_{jN}^{k-1})^3}]$$

$$+9.482[\frac{x_{li}^{k-2}-x_{lj}^{k-2}}{(r_{ij}^{k-2})^3}+\frac{x_{lj}^{k-2}}{(r_{jN}^{k-2})^3}]-6.798[\frac{x_{li}^{k-3}-x_{lj}^{k-3}}{(r_{ij}^{k-3})^3}+\frac{x_{lj}^{k-3}}{(r_{jN}^{k-3})^3}]$$

$$+2.627[\frac{x_{li}^{k-4}-x_{lj}^{k-4}}{(r_{ij}^{k-4})^3}+\frac{x_{lj}^{k-4}}{(r_{jN}^{k-4})^3}]-0.425[\frac{x_{li}^{k-5}-x_{lj}^{k-5}}{(r_{ij}^{k-5})^3}+\frac{x_{lj}^{k-5}}{(r_{jN}^{k-5})^3}]\}),$$

where $r_{iN}^{p}=[\sum_{l=1}^{3}(x_{li}^{p})^2]^{1/2}$,

$$r_{ij}^{p}=[\sum_{l=1}^{3}(x_{li}^{p}-x_{lj}^{p})^2]^{1/2}\quad(p=k-5(1)k);$$

4° for l=1,2,3 and i=1(1)N-1:

$\bar{x}_{li}^{k+1}:=x_{li}^{k+1}$;

$\bar{v}_{li}^{k+1}:=v_{li}^{k+1}$;

$$w_{li} := 6.5112v_{li}^{k} - 4.6461v_{li}^{k-1} + 3.7504v_{li}^{k-2} - 2.0211v_{li}^{k-3} + 0.6312v_{li}^{k-4} - 0.0863v_{li}^{k-5};$$

$$z_{li} := (m_N+m_i)\left[6.5112\frac{x_{li}^{k}}{(r_{iN}^{k})^3} - 4.6461\frac{x_{li}^{k-1}}{(r_{iN}^{k-1})^3} + 3.7504\frac{x_{li}^{k-2}}{(r_{iN}^{k-2})^3} - 2.0211\frac{x_{li}^{k-3}}{(r_{iN}^{k-3})^3} + 0.6312\frac{x_{li}^{k-4}}{(r_{iN}^{k-4})^3} - 0.0863\frac{x_{li}^{k-5}}{(r_{iN}^{k-5})^3}\right]$$

$$+ \sum_{\substack{j=1\\ j\neq i}}^{N-1} m_j\left\{6.5112\left[\frac{x_{li}^{k}-x_{lj}^{k}}{(r_{ij}^{k})^3} + \frac{x_{lj}^{k}}{(r_{jN}^{k})^3}\right] - 4.6461\left[\frac{x_{li}^{k-1}-x_{lj}^{k-1}}{(r_{ij}^{k-1})^3} + \frac{x_{lj}^{k-1}}{(r_{jN}^{k-1})^3}\right]\right.$$

$$+3.7504\left[\frac{x_{li}^{k-2}-x_{lj}^{k-2}}{(r_{ij}^{k-2})^3} + \frac{x_{lj}^{k-2}}{(r_{jN}^{k-2})^3}\right]$$

$$-2.0211\left[\frac{x_{li}^{k-3}-x_{lj}^{k-3}}{(r_{ij}^{k-3})^3} + \frac{x_{lj}^{k-3}}{(r_{jN}^{k-3})^3}\right] + 0.6312\left[\frac{x_{li}^{k-4}-x_{lj}^{k-4}}{(r_{ij}^{k-4})^3} + \frac{x_{lj}^{k-4}}{(r_{jN}^{k-4})^3}\right]$$

$$\left.-0.0863\left[\frac{x_{li}^{k-5}-x_{lj}^{k-5}}{(r_{ij}^{k-5})^3} + \frac{x_{lj}^{k-5}}{(r_{jN}^{k-5})^3}\right]\right\},$$

where $r_{iN}^{p} = [\sum_{l=1}^{3} (x_{li}^{p})^2]^{1/2}$,

$r_{ij}^{p} = [\sum_{l=1}^{3} (x_{li}^{p} - x_{lj}^{p})^2]^{1/2}$ $(p=k-5(1)k)$;

6° for $l=1,2,3$ and $i=1(1)N-1$:

$$v_{li}^{k+1} := v_{li}^{k} - \frac{\Delta t\times G}{6.0480}\left(1.9087\left\{(m_N+m_i)\frac{x_{li}^{k+1}}{(r_{iN}^{k+1})^3} + \sum_{\substack{j=1\\ j\neq i}}^{N-1} m_j\left[\frac{x_{li}^{k+1}-x_{lj}^{k+1}}{(r_{ij}^{k+1})^3} + \frac{x_{lj}^{k+1}}{(r_{jN}^{k+1})^3}\right]\right\} + z_{li}\right),$$

where $r_{iN}^{k+1} = [\sum_{l=1}^{3} (x_{li}^{k+1})^2]^{1/2}$,

$r_{ij}^{k+1} = [\sum_{l=1}^{3} (x_{li}^{k+1} - x_{lj}^{k+1})^2]^{1/2}$.

Algorithm 16. The Butcher method of sixth order (compare Algorithm 6)

$1^o, 2^o, 4^o, 6^o$ - as in Algorithm 6, but with $i=1(1)N-1$;
3° for $l=1,2,3$ and $i=1(1)N-1$:

$$w_{li} := v_{li}^{k-1} - \frac{\Delta t \times G}{8}\Big((m_N+m_i)\Big[9\,\frac{x_{li}^k}{(r_{iN}^k)^3} + 3\,\frac{x_{li}^{k-1}}{(r_{li}^{k-1})^3}\Big] + \sum_{\substack{j=1\\ j\neq i}}^{N-1} m_j\Big\{9\Big[\frac{x_{li}^k - x_{lj}^k}{(r_{ij}^k)^3} + \frac{x_{lj}^k}{(r_{jN}^k)^3}\Big] + 3\Big[\frac{x_{li}^{k-1} - x_{lj}^{k-1}}{(r_{ij}^{k-1})^3} + \frac{x_{lj}^{k-1}}{(r_{jN}^{k-1})^3}\Big]\Big\}\Big),$$

where $r_{iN}^p = \Big[\sum_{l=1}^{3} (x_{li}^p)^2\Big]^{1/2}$,

$$r_{ij}^p = \Big[\sum_{l=1}^{3} (x_{li}^p - x_{lj}^p)^2\Big]^{1/2} \quad (p=k-1,k);$$

$$z_{li} := x_{li}^{k-1} + \frac{\Delta t}{8}(9v_{li}^k + 3v_{li}^{k-1});$$

5° for $l=1,2,3$ and $i=1(1)N-1$:

$$x_{li}^{k+1} := \frac{1}{3.1}(3.2x_{li}^k - 0.1x_{li}^{k-1}) + \frac{\Delta t}{9.3}\Big[6.4w_{li} + 9.6v_{li}^k - 7.0v_{li}^{k-1} \times \Delta t \times G\Big((m_N+m_i)\Big[3.2\,\frac{z_{li}}{\bar{r}_{iN}^3} - 6.0\,\frac{x_{li}^k}{(r_{iN}^k)^3} - 2.6\,\frac{x_{li}^{k-1}}{(r_{iN}^{k-1})^3}\Big] + \sum_{\substack{j=1\\ j\neq i}}^{N-1} m_j\Big\{3.2\Big(\frac{z_{li}-z_{lj}}{\bar{r}_{ij}^3} + \frac{z_{lj}}{\bar{r}_{jN}^3}\Big) - 6.0\Big[\frac{x_{li}^k - x_{lj}^k}{(r_{ij}^k)^3} + \frac{x_{lj}^k}{(r_{jN}^k)^3}\Big] - 2.6\Big[\frac{x_{li}^{k-1} - x_{lj}^{k-1}}{(r_{ij}^{k-1})^3} + \frac{x_{lj}^{k-1}}{(r_{jN}^{k-1})^3}\Big]\Big\}\Big)\Big];$$

$$v_{li}^{k+1} := \frac{1}{3.1}(3.2v_{li}^k - 0.1v_{li}^{k-1}) - \frac{\Delta t \times G}{9.3}\Big((m_N+m_i)\Big[6.4\,\frac{z_{li}}{\bar{r}_{iN}^3} + 1.2\,\frac{x_{li}^k}{(r_{iN}^k)^3} - 0.1\,\frac{x_{li}^{k-1}}{(r_{iN}^{k-1})^3} + 1.5\,\frac{\tilde{z}_{li}}{\tilde{r}_{iN}^3}$$

$$+\sum_{\substack{j=1\\j\neq i}}^{N-1} m_j\{6.4(\frac{z_{li}-z_{lj}}{\bar{r}_{ij}^3}+\frac{z_{lj}}{\bar{r}_{jN}^3})+1.2[\frac{x_{li}^k-x_{lj}^k}{(r_{ij}^k)^3}+\frac{x_{lj}^k}{(r_{jN}^k)^3}]$$

$$-0.1[\frac{x_{li}^{k-1}-x_{lj}^{k-1}}{(r_{ij}^{k-1})^3}+\frac{x_{lj}^{k-1}}{(r_{jN}^{k-1})^3}]$$

$$+1.5(\frac{\tilde{z}_{li}-\tilde{z}_{lj}}{\tilde{r}_{ij}^3}+\frac{\tilde{z}_{lj}}{\tilde{r}_{jN}^3})\}),$$

where $r_{iN}^p=[\sum_{l=1}^{3}(x_{li}^p)^2]^{1/2}$, $r_{ij}^p=[\sum_{l=1}^{3}(x_{li}^p-x_{lj}^p)^2]^{1/2}$,

$\bar{r}_{iN}=(\sum_{l=1}^{3} z_{li}^2)^{1/2}$, $\bar{r}_{ij}=[\sum_{l=1}^{3}(z_{li}-z_{lj})^2]^{1/2}$,

$\tilde{r}_{iN}=(\sum_{l=1}^{3} \tilde{z}_{li}^2)^{1/2}$, $\tilde{r}_{ij}=[\sum_{l=1}^{3}(\tilde{z}_{li}-\tilde{z}_{lj})^2]^{1/2}$

$(p=k-1,k)$.

<u>Algorithm 17.</u> Rational extrapolation with an automatic step size correction (compare Algorithms 7 and 8)

$1^o,2^o,6^o,8^o$-10° - as in Algorithm 7, but with $i=1(1)N-1$ and $p=1(1)6N-6$;

3° for $p=1(1)6N-6$:

$T_{s,-1,p}:=0$;

$$\beta_s:=2\times\begin{cases}1, & \text{if } s=0,\\ 2^{(s+1)/2}, & \text{for } s \text{ odd},\\ 3\times 2^{(s-2)/2}, & \text{for } s \text{ even};\end{cases}$$

$\varkappa:=0$;

for $l=1,2,3$ and $i=1(1)N-1$:

$$z_{li}:=y_{li}+\frac{h_o}{\beta_s}u_{li};$$

$$w_{li}:=u_{li}-\frac{h_oG}{\beta_s}[(m_N+m_i)\frac{y_{li}}{\bar{r}_{iN}^3}+\sum_{\substack{j=1\\j\neq i}}^{N-1} m_j(\frac{y_{li}-y_{lj}}{\bar{r}_{ij}^3}+\frac{y_{lj}}{\bar{r}_{jN}^3})],$$

where $\bar{r}_{iN}=(\sum_{l=1}^{3} y_{li}^2)^{1/2}$, $\bar{r}_{ij}=[\sum_{l=1}^{3}(y_{li}-y_{lj})^2]^{1/2}$;

$\varkappa:=\max(\varkappa,|z_{li}|,|y_{li}|,|w_{li}|,|u_{li}|)$;

4° for $p=2(1)\beta_s-1$:

for $l=1,2,3$ and $i=1(1)N-1$:

$a_{li}:=z_{li}$;

$b_{li}:=w_{li}$;

$$w_{li}:=u_{li}-\frac{2h_oG}{\beta_s}[(m_N+m_i)\frac{a_{li}}{\tilde{r}^3_{iN}}+\sum_{\substack{j=1\\j\neq i}}^{N-1} m_j(\frac{a_{li}-a_{lj}}{\tilde{r}^3_{ij}}+\frac{a_{lj}}{\tilde{r}^3_{jN}})],$$

where $\tilde{r}_{iN}=(\sum_{l=1}^{3} a^2_{li})^{1/2}$, $\tilde{r}_{ij}=[\sum_{l=1}^{3}(a_{li}-a_{lj})^2]^{1/2}$;

$\varkappa:=\max(\varkappa,|z_{li}|,|w_{li}|)$;

for l=1,2,3 and i=1(1)N-1:

$y_{li}:=a_{li}$;

$u_{li}:=b_{li}$;

5° for l=1,2,3 and i=1(1)N-1:

p:=(l-1)(N-1)+i;

$$T_{s0p}:=\frac{1}{2}(z_{li}+y_{li})+\frac{h_o}{\beta_s}\{w_{li}+\frac{u_{li}}{2}-\frac{h_oG}{\beta_s}[(m_N+m_i)\frac{z_{li}}{\bar{r}^3_{iN}}+\sum_{\substack{j=1\\j\neq i}}^{N-1} m_j(\frac{z_{li}-z_{lj}}{\bar{r}^3_{ij}}+\frac{z_{lj}}{\bar{r}^3_{jN}})]\};$$

$$T_{s0,p+3N-3}:=\frac{1}{2}(w_{li}+u_{li})-\frac{h_oG}{\beta_s}\{(m_N+m_i)(\frac{z_{li}}{\bar{r}^3_{iN}}+\frac{y_{li}+\frac{2h_o}{\beta_s}w_{li}}{2\tilde{r}^3_{iN}})+\sum_{\substack{j=1\\j\neq i}}^{N-1} m_j[\frac{z_{li}-z_{lj}}{\bar{r}^3_{ij}}+\frac{z_{lj}}{\bar{r}^3_{jN}}+\frac{y_{lj}+\frac{2h_o}{\beta_s}w_{lj}}{2\tilde{r}^3_{jN}}+\frac{y_{li}-y_{lj}+\frac{2h_o}{\beta_s}(w_{li}-w_{lj})}{2\tilde{r}^3_{ij}}]\},$$

where $\bar{r}_{iN}=(\sum_{l=1}^{3} z^2_{li})^{1/2}$, $\bar{r}_{ij}=[\sum_{l=1}^{3}(z_{li}-z_{lj})^2]^{1/2}$

$$\tilde{r}_{iN}=[\sum_{l=1}^{3}(y_{li}+\frac{2h_o}{\beta_s}w_{li})^2]^{1/2},$$

$$\tilde{r}_{ij}=\{\sum_{l=1}^{3}[y_{li}-y_{lj}+\frac{2h_o}{\beta_s}(w_{li}-w_{lj})]^2\}^{1/2}.$$

7° as in Algorithm 8, but with p=1(1)6N-6.

The above algorithms used in our model problem of relative

motion of two bodies (see p.144) give the results presented in Tables XXVIII-XXXI. Let us note that the values of d_3 and d_4 are not constant (see Fig.22). In Table XXXII we present the solution of the six-body problem relative motion by the rational extrapolation method. In the Runge-Kutta and rational extrapolation method we have assumed the same admissible discretization error as in Sect.3.2.

Table XXVIII. Solution of the problem of relative motion of two bodies by the fourth order method of Runge--Kutta (Δt=0.01)

ν	l	x_l	v_l	characteristics constants of motion
20	1	0.309016957331	-5.975664650413	α=16, β=3
	2	0.951056518886	1.941610822396	Δ=0.000001%
				d_3=6.283185489109
				d_4=-19.73920997913
60	1	-0.809016857631	3.693164830810	α=16, β=3
	2	-0.587785398556	-5.083203252590	Δ=0.000008%
				d_3=6.283185484122
				d_4=-19.73921001046
100	1	0.999999995915	-0.000001987183	α=16, β=3
	2	0.000000315996	6.283185504564	Δ=0.000013%
				d_3=6.283185478895
				d_4=-19.73921004331
				CPU time 0.69 sec

Table XXIX. Solution of the problem of relative motion of two bodies by the predictor-corrector method

ν	l	x_l	v_l	characteristics constants of motion
20	1	0.309016959277	-5.975664576875	α=4, β=1
	2	0.951056527747	1.941610875517	Δ=0.1×10⁻⁷%
				d_3=6.283185492318
				d_4=-19.73920995897
60	1	-0.809016930389	3.693164324744	α=4, β=1
	2	-0.587785341463	-5.083203435574	Δ=1.2×10⁻⁷%
				d_3=6.283185493692
				d_4=-19.73920995034
100	1	1.000000001081	-0.000001130677	α=4, β=1
	2	0.000000179945	6.283185488273	Δ=5.5×10⁻⁷%
				d_3=6.283185495065
				d_4=-19.73920994170
				CPU time 0.51 sec

Let us pay attention to the very good approximation of the exact solution by the seventh order predictor-corrector method.

Table XXX. Solution of the problem of relative motion of two bodies by the Butcher method ($\Delta t=0.01$)

ν	l	x_l	v_l	characteristics constants of motion
20	1	0.309016931494	-5.975664745974	$\alpha=4$, $\beta=1$
	2	0.951056529056	1.941610941490	$\Delta=0.000003\%$
				$d_3=6.283185627401$
				$d_4=-19.73920911021$
60	1	-0.809017143547	3.693162741666	$\alpha=4$, $\beta=1$
	2	-0.587785308954	-5.083203913632	$\Delta=0.000021\%$
				$d_3=6.283185913398$
				$d_4=-19.73920731324$
100	1	1.000000228565	0.000005165360	$\alpha=4$, $\beta=1$
	2	-0.000000821720	6.283184763273	$\Delta=0.000118\%$
				$d_3=6.283186199394$
				$d_4=-19.73920551628$

CPU time 0.42 sec

Table XXXI. Solution of the problem of relative motion of two bodies by the rational extrapolation method ($\Delta t=0.01$)

ν	l	x_l	v_l	characteristics constants of motion
20	1	0.309016959250	-5.975664576794	$\alpha=510$, $\beta=29$
	2	0.951056527705	1.941610875112	$\Delta=0.5\times10^{-9}\%$
				$d_3=6.283185491809$
				$d_4=-19.73920996216$
60	1	-0.809016929241	3.693164332780	$\alpha=452$, $\beta=31$
	2	-0.587785341926	-5.083203432494	$\Delta=2.9\times10^{-9}\%$
				$d_3=6.283185491795$
				$d_4=-19.73920996230$
100	1	0.999999999993	-0.000001160481	$\alpha=556$, $\beta=28$
	2	0.000000184706	6.283185491849	$\Delta=8.5\times10^{-9}\%$
				$d_3=6.283185491804$
				$d_4=-19.73920996219$

CPU time 13.01 sec

Table XXXII. Solution of the problem of relative motion of six bodies by the rational extrapolation method (the data at the initial moment as in Table XIV)

ν	i	l	x_{li}	v_{li}	characteristics
40	1	1	0.797521441667	-3.846101036778	α=251, β=13
		2	0.591449510260	5.024740543867	
		3	0.000000160695	-0.000000728301	
	2	1	0.316824631387	-2.785348915514	
		2	5.106401379870	0.299586598532	
		3	-0.072774778587	-0.049045678446	
	3	1	-7.365357695787	-1.332865525084	
		2	5.508504902735	-1.636667197143	
		3	0.198596806639	0.081412918809	
	4	1	-13.812298449420	0.948258532378	
		2	-12.453669729551	-1.133988815602	
		3	0.133077332993	-0.016490524103	
	5	1	-7.516422378007	1.100507463741	
		2	-29.329500242376	-0.278288905123	
		3	0.776227869511	-0.019769186355	
70	1	1	-0.834758934335	-3.464409188072	α=251, β=13
		2	0.528523506197	-5.332190271517	
		3	-0.000000677580	-0.000002976614	
	2	1	-0.519299401452	-2.776636625845	
		2	5.128748117673	-0.149232366823	
		3	-0.086467295480	-0.042064369531	
	3	1	-7.748131220314	-1.218135002996	
		2	5.005275400932	-1.716803261165	
		3	0.222550245336	0.078220678335	
	4	1	-13.524030419818	0.973451193250	
		2	-12.790467700655	-1.111235845250	
		3	0.128094733656	-0.016725021065	
	5	1	-7.185793202208	1.103686100432	
		2	-29.411180025646	-0.266234844109	
		3	0.770248391119	-0.020093436116	
100	1	1	-0.259051570427	5.973401148516	α=174, β=12
		2	-0.981127566209	-1.626983735616	
		3	-0.000000193118	0.000006026661	
	2	1	-1.342061246468	-2.697023380996	
		2	5.018268545686	-0.584039886931	
		3	-0.097921416990	-0.034163398501	
	3	1	-8.095781615869	-1.098837325924	
		2	4.479275908670	-1.788418351526	
		3	0.245496694273	0.074703277689	
	4	1	-13.228278190767	0.998135748776	
		2	-13.120349759234	-1.087874776089	
		3	0.123043383881	-0.016948883217	
	5	1	-6.854214876500	1.106819598396	
		2	-29.489231108176	-0.254087045637	
		3	0.764172138894	-0.020414375539	

CPU time 141.67 sec

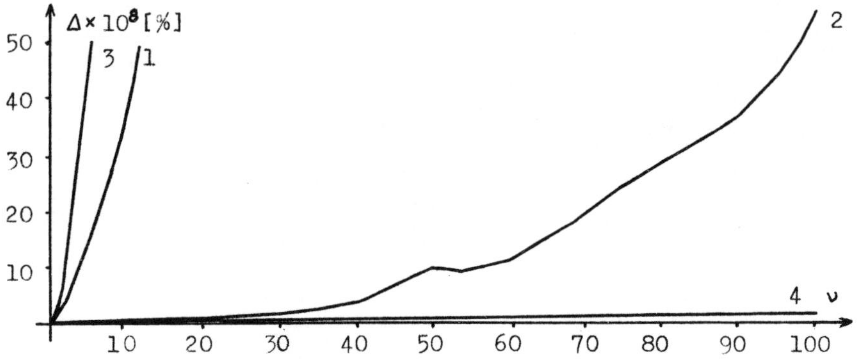

Fig.21. Mean relative errors in the problem of relative motion of two bodies: 1 - the Runge-Kutta method, 2 - the predictor-corrector method, 3 - the method of Butcher, 4 - rational extrapolation

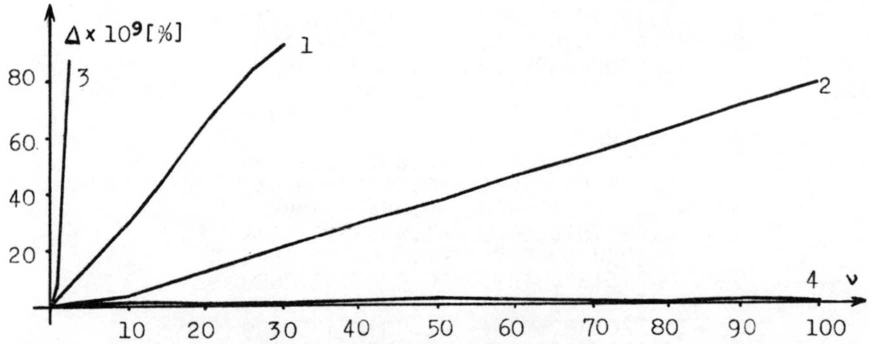

Fig.22. Mean relative errors of constants of motion in the problem of relative motion of two bodies ($\Delta = (|d_3 - \bar{d}_3| / |\bar{d}_3| + |d_4 - \bar{d}_4| / |\bar{d}_4|)/2$, $\bar{d}$ denotes the exact value)

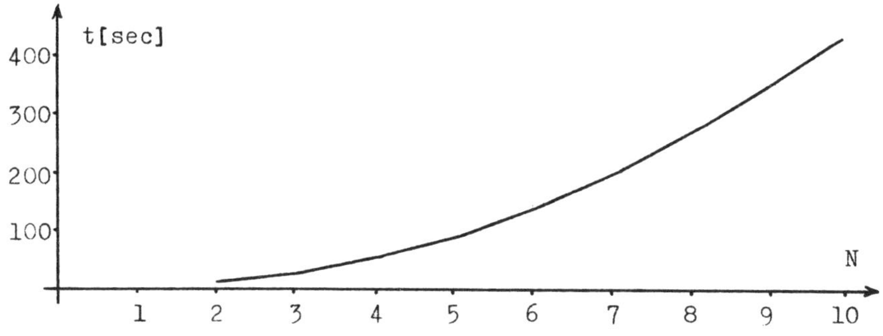

Fig.23. Increase of computational time with respect to the number of bodies for the rational extrapolation method

3.3. THE APPLICATION OF DISCRETE MECHANICS FORMULAS

Let us now give some thought to the form of Greenspan's discrete mechanics formulas (2.3.1),(2.3.4)-(2.3.5) in the frame of reference with an origin in some material point of mass m_N. It is easy to show that transforming an origin of an inertial frame to this point, the form of the equations (2.3.4)-(2.3.5) does not change. However, the form of gravitational force (2.3.1) will be changed. If we do analogous transformations as in Sect.3.1, then from (2.3.1) we get that in relative motion

$$F_{li}^k = -Gm_i\{(m_N+m_i)\frac{x_{li}^{k+1}+x_{li}^k}{r_{iN}^{k+1}r_{iN}^k(r_{iN}^{k+1}+r_{iN}^k)} + \sum_{\substack{j=1\\j\neq i}}^{N-1} m_j[\frac{x_{li}^{k+1}+x_{li}^k-x_{lj}^{k+1}-x_{lj}^k}{r_{ij}^{k+1}r_{ij}^k(r_{ij}^{k+1}+r_{ij}^k)} + \frac{x_{lj}^{k+1}+x_{lj}^k}{r_{jN}^{k+1}r_{jN}^k(r_{jN}^{k+1}+r_{jN}^k)}]\}; \tag{3.3.1}$$

$$l=1,2,3;\ i=1(1)N-1;\ k=0,1,2,\ldots;$$

where

$$r_{iN}^s=[\sum_{l=1}^{3}(x_{li}^s)^2]^{1/2},\quad r_{ij}^s=[\sum_{l=1}^{3}(x_{li}^s-x_{lj}^s)^2]^{1/2};$$
$$s=k \text{ or } s=k+1.$$

In the case of relative motion of two bodies the formulas (2.3.4)-(2.3.5) and (3.3.1) yield the following system of equations

$$\frac{v_l^{k+1}+v_l^k}{2} = \frac{x_l^{k+1}-x_l^k}{\Delta t},$$
$$\frac{F_l^k}{m_1} = \frac{v_l^{k+1}-v_l^k}{\Delta t}, \tag{3.3.2}$$
$$F_l^k = -Gm_1(m_1+m_2)\frac{x_l^{k+1}+x_l^k}{r^{k+1}r^k(r^{k+1}+r^k)};$$
$$l=1,2,3;\ k=0,1,2,\ldots;$$

where

$$r^s=[\sum_{l=1}^{3}(x_l^s)^2]^{1/2}\quad (s=k \text{ or } s=k+1).$$

We can show that the equations (3.3.2) conserve the constants of motion given by (3.1.6) and (3.1.8). We have

<u>Theorem 3.1.</u> If the relative motion of two bodies is described by the discrete equations (3.3.2), then for each $k=0,1,2,\ldots$

$$x_p^k v_q^k - x_q^k v_p^k = d_l, \tag{3.3.3}$$

where $l=1$ for $(p,q)=(2,3)$, $l=2$ for $(p,q)=(3,1)$, $l=3$ for $(p,q)=(1,2)$, d_l=const and

$$\frac{1}{2}\sum_{l=1}^{3}(v_l^k)^2 - \frac{G(m_1+m_2)}{r^k} = d_4, \tag{3.3.4}$$

where d_4=const.

Proof. The equations (3.3.3) are equivalent to the relation

$$x^{k+1}\times v^{k+1} - x^k\times v^k = 0,$$

which should hold for each $k=0,1,2,\ldots$ and where $\times$ denotes a vector product. From (3.3.2) we have

$$x^{k+1}\times v^{k+1}-x^k\times v^k = \Delta t x^{k+1}\times(\frac{v^{k+1}-v^k}{\Delta t})+(x^{k+1}-x^k)\times v^k$$
$$= \frac{\Delta t}{m_1}x^{k+1}\times F^k - \frac{\Delta t}{2m_1}(x^{k+1}-x^k)\times F^k$$
$$= \frac{\Delta t}{2m_1}(x^{k+1}+x^k)\times F^k$$

$$= -\frac{G}{2}(m_1+m_2)\frac{(x^{k+1}+x^k)\times(x^{k+1}+x^k)}{r^{k+1}r^k(r^{k+1}+r^k)} = 0.$$

In order to prove (3.3.4) we show that for each k=0,1,2,...

$$\frac{1}{2}\sum_{l=1}^{3}[(v_l^{k+1})^2-(v_l^k)^2] = G(m_1+m_2)\left(\frac{1}{r^{k+1}}-\frac{1}{r^k}\right).$$

From (3.3.2) we get

$$\frac{1}{2}\sum_{l=1}^{3}[(v_l^{k+1})^2-(v_l^k)^2] = \Delta t\left(\frac{v^{k+1}-v^k}{\Delta t}\right)\circ\left(\frac{v^{k+1}+v^k}{2}\right)$$

$$= \frac{1}{m_1}F^k\circ(x^{k+1}-x^k)$$

$$= -G(m_1+m_2)\frac{(x^{k+1}+x^k)\circ(x^{k+1}-x^k)}{r^{k+1}r^k(r^{k+1}+r^k)}$$

$$= -G(m_1+m_2)\frac{(r^{k+1})^2-(r^k)^2}{r^{k+1}r^k(r^{k+1}+r^k)} = G(m_1+m_2)\left(\frac{1}{r^{k+1}}-\frac{1}{r^k}\right).$$

Let us recall to mind that conventional numerical methods applied to the problem of relative motion of two bodies do not have the properties (3.3.3) and (3.3.4) (see the previous section). Of course, in the case of two bodies the discrete equations (3.3.2) are not of great importance, because this problem can be solved analytically. However, for N>2 the formulas of discrete mechanics for relative motion have much better numerical properties than conventional methods of similar accuracy order. We will see it further on in this section. Now, we briefly consider the problem of stability and convergence of the method (2.3.4)-(2.3.5), (3.3.1).

Without loss of generality let us assume, as in Sect.2.3.2, that the method (2.3.4)-(2.3.5),(3.3.1) is used to solve the problem of relative motion of N bodies in the interval [0,1] and let $\Delta t=1/n$. Let us denote $y_p\equiv x_{li}$, $y_{p+3N-3}\equiv v_{li}$, where $l=[\frac{p-1}{N-1}]+1$, $i=p-[\frac{p-1}{N-1}](N-1)$ ([.] means an entire part of), $f_p(y)\equiv F_{li}/m_i$, where

$$F_{li} = -Gm_i\left[(m_N+m_i)\frac{x_{li}}{r_{iN}^3} + \sum_{\substack{j=1\\j\neq i}}^{N-1} m_j\left(\frac{x_{li}-x_{lj}}{r_{ij}^3}+\frac{x_{lj}}{r_{jN}^3}\right)\right]. \quad (3.3.5)$$

Let $E=\times_{6N-6}\, C^{(1)}[0,1]$, $E^{o}=\times_{6N-6}\, R\times\times_{6N-6}\, C^{(1)}[0,1]$, where the norms in E and E^o are determined by the formulas (1.1.5). The problem (3.1.4) we can write now in the form

$$\Phi y = \begin{cases} y_p(0)-z_p^o, \\ y_{p+3N-3}(0)-z_{p+3N-3}^o, \\ y_p'-y_{p+3N-3}, \\ y_{p+3N-3}'-f_p(y); \quad p=1(1)3N-3; \end{cases} \tag{3.3.6}$$

where $z_p^o \equiv x_{1i}^o$, $z_{p+3N-3}^o \equiv v_{1i}^o$ and $y \epsilon E^o$ for $y \epsilon E$. The problem (3.3.6) corresponds to the discrete initial value problem given by

$$\Phi_n \eta(\tfrac{\nu}{n}) = \begin{cases} \eta_p(0)-z_p^o, \text{ for } \nu=0, \\ \eta_{p+3N-3}(0)-z_{p+3N-3}^o, \text{ for } \nu=0, \\ \dfrac{\eta_p(\frac{\nu}{n})-\eta_p(\frac{\nu-1}{n})}{1/n} \\ \quad -[\eta_{p+3N-3}(\frac{\nu}{n})+\eta_{p+3N-3}(\frac{\nu-1}{n})], \\ \qquad \text{for } \nu=1(1)n, \\ \dfrac{\eta_{p+3N-3}(\frac{\nu}{n})-\eta_{p+3N-3}(\frac{\nu-1}{n})}{1/n} \\ \quad -F_p[\eta(\frac{\nu}{n}),\eta(\frac{\nu-1}{n})], \quad \text{for } \nu=1(1)n; \end{cases} \tag{3.3.7}$$

$$p=1(1)3N-3;$$

where the function $F_p \equiv F_{1i}^{\nu-1}/m_i$ is determined by (3.3.1) and $\Phi_n \eta \epsilon E_n^o$ for $\eta \epsilon E_n$ (the norms in E_n and E_n^o are given by (1.2.1)).

Doing analogously as in Sect.2.3.2, we can prove the following theorems and lemmas (we omit the proofs):

Theorem 3.2. The method (3.3.7) is consistent with the initial value problem (3.3.6).

Lemma 3.1. If the function

$$F_p[x(t),y(t)] \equiv F_{1i}[x(t),y(t)]$$
$$= -G\{(m_N+m_i)\frac{y_{1i}+x_{1i}}{r_{iN}R_{iN}(r_{iN}+R_{iN})}$$
$$+\sum_{\substack{j=1\\j\neq i}}^{N-1} m_j[\frac{y_{1i}+x_{1i}-y_{1j}-x_{1j}}{r_{ij}R_{ij}(r_{ij}+R_{ij})}+\frac{y_{1j}+x_{1j}}{r_{jN}R_{jN}(r_{jN}+R_{jN})}]\},$$

where

$$r_{iN}=(\sum_{l=1}^{3} x_{li}^2)^{1/2},\quad r_{ij}=[\sum_{l=1}^{3}(x_{li}-x_{lj})^2]^{1/2},$$

$$R_{iN}=(\sum_{l=1}^{3} y_{li}^2)^{1/2},\quad R_{ij}=[\sum_{l=1}^{3}(y_{li}-y_{lj})^2]^{1/2},$$

is such that $y(t)=x(t+h)$, where $t+h\epsilon[0,1]$ if $t\epsilon[0,1]$ and there exist the constants (2.3.23), then for arbitrary $p,q=1(1)3N-3$

$$\left.\begin{array}{l} \left|\dfrac{\partial F_p}{\partial y_q}\right| \\ \left|\dfrac{\partial F_p}{\partial x_q}\right| \end{array}\right\} \le \frac{GMN}{r^3}(1+\frac{3R}{r}).$$

Lemma 3.2. If there exist the constants (2.3.23), then there exists $\tilde{M}\ge 3N-3$ such that

$$|F_p(\bar{x},\bar{y})-F_p(x,y)|\le \tilde{M}\sum_{q=1}^{3N-3}(|\bar{x}_q-x_q|+|\bar{y}_q-y_q|);$$
$$p=1(1)3N-3.$$

Theorem 3.3. Let us assume the existance of the constants (2.3.23). Then:

(i) the method (3.3.7) is of the first order;

(ii) for $n>\tilde{M}(3N-3)$, where $\tilde{M}$ denotes a constant occuring in Lemma 3.2, the method (3.3.7) is stable on the initial value problem (3.3.6), where the constant S from Def.1.5 has the value

$$S = e^{(3N-3)\tilde{M}}\{\frac{(3N-3)([\tilde{M}]-\tilde{M})+1}{(3N-3)[\tilde{M}]+1}\}^{-(3N-3)[\tilde{M}]-1};$$

(iii) the method (3.3.7) is convergent on the initial value problem (3.3.6).

Let us return to the previous notation of our method (i.e. to the formulas (2.3.4)-(2.3.5),(3.3.1)). An algorithm with an automatic step size correction for this method is not much different from the Algorithm 9. Therefore, the below description contains only essential modifications.

Algorithm 18.

1°-3°,5°-8°,10°,11° - as in Algorithm 9, but the summation with respect to i ought to be done from 1 to N-1;

4° for $l=1,2,3$ and $i=1(1)N-1$:

$$\Phi:=x_{li}^{k+1}-x_{li}^{k}-Hv_{li}^{k}$$

$$+\frac{H^2G}{2}\{(m_N+m_i)\frac{x_{li}^{k+1}+x_{li}^{k}}{r_{iN}^{k+1}r_{iN}^{k}(r_{iN}^{k+1}+r_{iN}^{k})}$$

$$+\sum_{\substack{j=1\\j\neq i}}^{N-1} m_j[\frac{x_{li}^{k+1}+x_{li}^{k}-x_{lj}^{k+1}-x_{lj}^{k}}{r_{ij}^{k+1}r_{ij}^{k}(r_{ij}^{k+1}+r_{ij}^{k})}+\frac{x_{lj}^{k+1}+x_{lj}^{k}}{r_{jN}^{k+1}r_{jN}^{k}(r_{jN}^{k+1}+r_{jN}^{k})}]\};$$

$$\partial\Phi:=1+\frac{H^2G}{2}\{\frac{m_N+m_i}{r_{iN}^{k+1}r_{iN}^{k}(r_{iN}^{k+1}+r_{iN}^{k})}[1-\frac{(x_{li}^{k+1}+x_{li}^{k})x_{li}^{k+1}}{r_{iN}^{k+1}}\times(\frac{1}{r_{iN}^{k+1}}+\frac{1}{r_{iN}^{k+1}+r_{iN}^{k}})]$$

$$+\sum_{\substack{j=1\\j\neq i}}^{N-1}\frac{m_j}{r_{ij}^{k+1}r_{ij}^{k}(r_{ij}^{k+1}+r_{ij}^{k})}\times[1-\frac{(x_{li}^{k+1}+x_{li}^{k}-x_{lj}^{k+1}-x_{lj}^{k})(x_{li}^{k+1}-x_{lj}^{k+1})}{r_{ij}^{k+1}}\times(\frac{1}{r_{ij}^{k+1}}+\frac{1}{r_{ij}^{k+1}+r_{ij}^{k}})]\},$$

where $r_{iN}^{s}=[\sum_{l=1}^{3}(x_{li}^{s})^2]^{1/2}$, $r_{ij}^{s}=[\sum_{l=1}^{3}(x_{li}^{s}-x_{lj}^{s})^2]^{1/2}$ (s=k or s=k+1);

$x_{li}^{k+1}:=x_{li}^{k+1}-\frac{\Phi}{\partial\Phi}$;

for l=1,2,3 and i=1(1)N-1:

$v_{li}^{k+1}:=\frac{2}{H}(x_{li}^{k+1}-x_{li}^{k})-v_{li}^{k}$;

9° for l=1,2,3 and i=1(1)N-1:

$$\Phi:=z_{li}-y_{li}-\frac{H}{2}w_{li}+\frac{H^2G}{8}\{(m_N+m_i)\frac{z_{li}+y_{li}}{\tilde{r}_{iN}r_{iN}(\tilde{r}_{iN}+r_{iN})}$$

$$+\sum_{\substack{j=1\\j\neq i}}^{N-1} m_j[\frac{z_{li}+y_{li}-z_{lj}-y_{lj}}{\tilde{r}_{ij}r_{ij}(\tilde{r}_{ij}+r_{ij})}+\frac{z_{lj}+y_{lj}}{\tilde{r}_{jN}r_{jN}(\tilde{r}_{jN}+r_{jN})}]\};$$

$$\partial\Phi:=1+\frac{H^2G}{8}\{\frac{m_N+m_i}{\tilde{r}_{iN}r_{iN}(\tilde{r}_{iN}+r_{iN})}[1-\frac{(z_{li}+y_{li})z_{li}}{\tilde{r}_{iN}}(\frac{1}{\tilde{r}_{iN}}+\frac{1}{\tilde{r}_{iN}+r_{iN}})]$$

$$+\sum_{\substack{j=1\\ j\neq i}}^{N-1} \frac{m_j}{\tilde{r}_{ij} r_{ij}(\tilde{r}_{ij}+r_{ij})}\Big[1-\frac{(z_{li}+y_{li}-z_{lj}-y_{lj})(z_{li}-z_{lj})}{\tilde{r}_{ij}} \times\Big(\frac{1}{\tilde{r}_{ij}}+\frac{1}{\tilde{r}_{ij}+r_{ij}}\Big)\Big],$$

where $\tilde{r}_{iN}=(\sum_{l=1}^{3} z_{li}^2)^{1/2}$, $\tilde{r}_{ij}=[\sum_{l=1}^{3}(z_{li}-z_{lj})^2]^{1/2}$,

$r_{iN}=(\sum_{l=1}^{3} y_{li}^2)^{1/2}$, $r_{ij}=[\sum_{l=1}^{3}(y_{li}-y_{lj})^2]^{1/2}$;

$z_{li}:=z_{li}-\frac{\Phi}{\partial\Phi}$;

for $l=1,2,3$ and $i=1(1)N-1$:

$u_{li}:=\frac{4}{H}(z_{li}-y_{li})+w_{li}$.

The application of the above algorithm to our model problem of relative motion of two bodies (see p.144) yields the results given in Table XXXIII. Let us note that, taking no account of rounding errors, we have d_3,d_4=const.

Table XXXIII. Solution of the problem of relative motion of two bodies by the discrete mechanics method ($\Delta t=0.01$, $\varepsilon=0.000000001$)

ν	l	x_l	v_l	characteristics constants of motion
20	1	0.309017040683	-5.975664410671	$\alpha=388$, $\beta=71$
	2	0.951056501227	1.941611386624	$\Delta=0.000008\%$
				$d_3=6.283185491772$
				$d_4=-19.73920996240$
60	1	-0.809017076617	3.693163057964	$\alpha=377$, $\beta=69$
	2	-0.587785138993	-5.083204359031	$\Delta=0.000025\%$
				$d_3=6.283185491738$
				$d_4=-19.73920996261$
100	1	0.999999999981	0.000001479363	$\alpha=399$, $\beta=73$
		-0.000000235451	6.283185491873	$\Delta=0.000042\%$
				$d_3=6.283185491751$
				$d_4=-19.73920996253$
				CPU time 11.06 sec

Remark! On notations - see p.69.

Table XXXIV. Solution of the problem of relative motion of six bodies by the discrete mechanics method (Δt=0.01, ε=0.000000001, the data at the initial moment as in Table XIV)

ν	i	l	x_{li}	v_{li}	characteristics
40	1	1	0.797521522608	-3.846100318850	α=377, β=69
		2	0.591449382229	5.024741173076	
		3	0.000000160695	-0.000000728301	
	2	1	0.316824631585	-2.785348916777	
		2	5.106401379172	0.299586595235	
		3	-0.072774778561	-0.049045678301	
	3	1	-7.365357694905	-1.332865520812	
		2	5.508504901987	-1.636667201275	
		3	0.198596806583	0.081412918536	
	4	1	-13.809012387980	0.948264210634	
		2	-12.453667327756	-1.133976795197	
		3	0.133077305786	-0.016490658362	
	5	1	-7.516422377180	1.100507467958	
		2	-29.329500230573	-0.278288843286	
		3	0.776227869461	-0.019769186609	
70	1	1	-0.834758800592	-3.464410567013	α=388 , β=71
		2	0.528523715383	-5.332189366224	
		3	-0.000000677580	-0.000002976615	
	2	1	-0.519299401216	-2.776638626879	
		2	5.128748115534	-0.149232373528	
		3	-0.086467295371	-0.042064369163	
	3	1	-7.748131217606	-1.218134995256	
		2	5.005275398489	-1.716803268560	
		3	0.222550245159	0.078220677821	
	4	1	-13.520742080719	0.973460628623	
		2	-12.790460335662	-1.111214774293	
		3	0.128094651957	-0.016725249109	
	5	1	-7.185793199590	1.103686108115	
		2	-29.411179987219	-0.266234728591	
		3	0.770248390960	-0.020093436592	
100	1	1	-0.259051962850	5.973400489581	α=377 , β=69
		2	-0.981127464395	-1.626986035028	
		3	-0.000000193119	0.000006026661	
	2	1	-1.342061245993	-2.697023380301	
		2	5.018268541114	-0.584039897156	
		3	-0.097921416735	-0.034163397933	
	3	1	-8.095781610302	-1.098837314848	
		2	4.479275903623	-1.788418361734	
		3	0.245496693907	0.074703276957	
	4	1	-13.224986510191	0.998148520632	
		2	-13.120334713613	-1.087844641995	
		3	0.123043220430	-0.016949199219	

Table XXXIV. (cont.)

ν	i	l	x_{li}	v_{li}	characteristics
	5	1	-6.854214871082	1.106819609322	
		2	-29.489231027675	-0.254086881897	
		3	0.764172138561	-0.020414376212	
					CPU time 269.52 sec

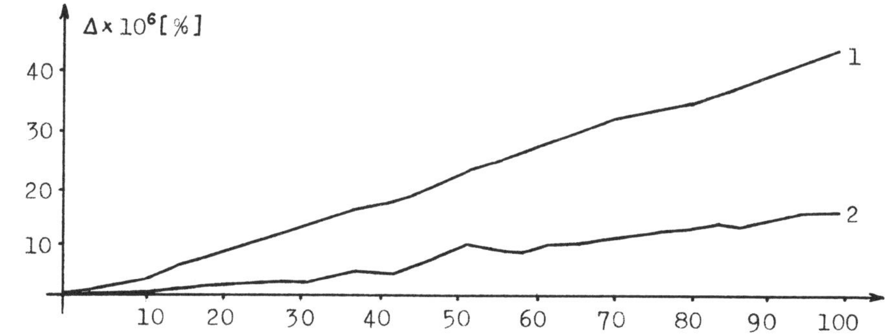

Fig.24. Mean relative errors for the problem of relative motion of two bodies: 1 - discrete mechanics of first order, 2 - the Runge-Kutta method of fourth order

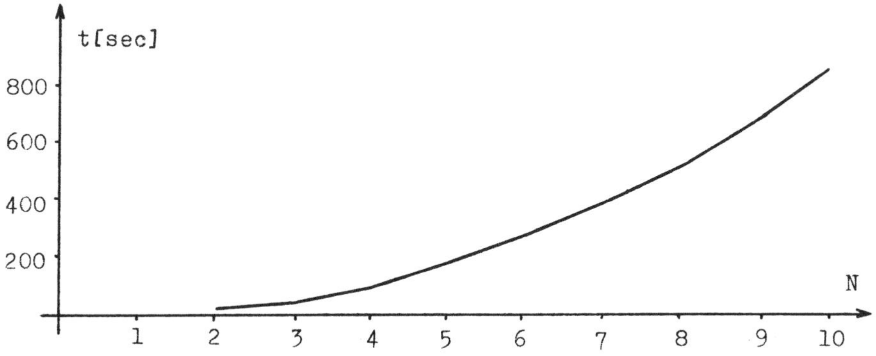

Fig.25. Increase of computational time with respect to the number of bodies

Chapter 4

Equations of Motion in a Rotating Frame

4.1. EQUATIONS OF MOTION

In practical applications the motion of the system of N material points is often described, especially when N=3, in a rotating frame of reference. The equations of motion in such a frame can be applied e.g. to the study of motion of a body with infinitely small mass nearby the equilibrium points or to the study of the moon's motion.

Let us assume that the system $x_1x_2x_3$ of rectangular coordinates rotates around the x_3-axis with the angular velocity θ and let us denote the axes of the rotating frame by ξ, η and ζ (see Fig.26). Between coordinates of an arbitrary point P in these two systems the following relationships hold

$$
\begin{aligned}
x_1 &= \xi \cos\theta - \eta \sin\theta, \\
x_2 &= \xi \sin\theta + \eta \cos\theta, \qquad (4.1.1) \\
x_3 &= \zeta.
\end{aligned}
$$

Differentiating two times the relations (4.1.1), we get

$$
\begin{aligned}
v_1 &\equiv \dot{x}_1 = (\dot{\xi}-\dot{\theta}\eta)\cos\theta - (\dot{\eta}+\dot{\theta}\xi)\sin\theta, \\
v_2 &\equiv \dot{x}_2 = (\dot{\xi}-\dot{\theta}\eta)\sin\theta + (\dot{\eta}+\dot{\theta}\xi)\cos\theta, \\
v_3 &\equiv \dot{x}_3 = \zeta, \qquad (4.1.2) \\
\dot{v}_1 &= (\ddot{\xi}-2\dot{\theta}\dot{\eta}-\dot{\theta}^2\xi-\ddot{\theta}\eta)\cos\theta - (\ddot{\eta}+2\dot{\theta}\dot{\xi}-\dot{\theta}^2\eta+\ddot{\theta}\xi)\sin\theta, \\
\dot{v}_2 &= (\ddot{\xi}-2\dot{\theta}\dot{\eta}-\dot{\theta}^2\xi-\ddot{\theta}\eta)\sin\theta + (\ddot{\eta}+2\dot{\theta}\dot{\xi}-\dot{\theta}^2\eta+\ddot{\theta}\xi)\cos\theta, \\
\dot{v}_3 &= \ddot{\zeta}.
\end{aligned}
$$

If we substitute (4.1.2) in the equations (2.1.1)-(2.1.2),

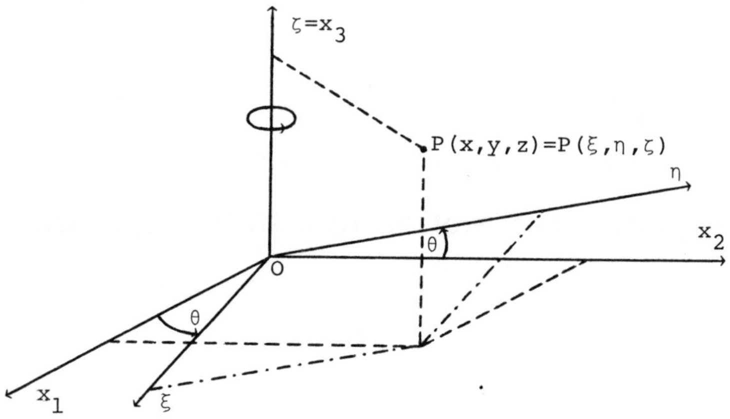

Fig.26. Frames of reference

and we equate the coefficients of 1, sin θ and cos θ, then for the motion of N bodies in the frame ξηζ we obtain the following equations

$$\begin{aligned} \ddot{\xi}_i - 2\dot{\theta}\dot{\eta}_i - \dot{\theta}^2\xi_i - \ddot{\theta}\eta_i &= F_{1i}/m_i, \\ \ddot{\eta}_i + 2\dot{\theta}\dot{\xi}_i - \dot{\theta}^2\eta_i + \ddot{\theta}\xi_i &= F_{2i}/m_i, \\ \ddot{\zeta}_i &= F_{3i}/m_i; \quad i=1(1)N; \end{aligned} \tag{4.1.3}$$

where

$$F_{li} = -Gm_i \sum_{\substack{j=1 \\ j\neq i}}^{N} m_j \frac{\alpha_{li}-\alpha_{lj}}{[(\xi_i-\xi_j)^2+(\eta_i-\eta_j)^2+(\zeta_i-\zeta_j)^2]^{3/2}}; \quad l=1,2,3; \tag{4.1.4}$$

and where $\alpha_{1i}=\xi_i$, $\alpha_{2i}=\eta_i$, $\alpha_{3i}=\zeta_i$.

For further simplicity it will be convenient to denote the axes of the frame ξηζ again by x_1, x_2 and x_3. If $\dot{\theta}=\phi=$const, i.e. an angular velocity of rotation of the frame is constant (it is often assumed in practical applications), then from (4.1.3)-(4.1.4) we get

$$\begin{aligned} \ddot{x}_{1i} - 2\phi\dot{x}_{2i} - \phi^2 x_{1i} &= F_{1i}/m_i, \\ \ddot{x}_{2i} + 2\phi\dot{x}_{1i} - \phi^2 x_{2i} &= F_{2i}/m_i, \\ \ddot{x}_{3i} &= F_{3i}/m_i; \quad i=1(1)N; \end{aligned} \tag{4.1.5}$$

where

$$F_{li} = -Gm_i \sum_{\substack{j=1\\ j\neq i}}^{N} m_j \frac{x_{li}-x_{lj}}{\left[\sum_{l=1}^{3} (x_{li}-x_{lj})^2\right]^{3/2}} . \tag{4.1.6}$$

Now, let us multiply the equations (4.1.5) by $2\dot{x}_{1i}$, $2\dot{x}_{2i}$ and $2\dot{x}_{3i}$ respectively, and then let us add them together and obtain the sum with respect to the index i. We have

$$2 \sum_{i=1}^{N} m_i [\ddot{x}_{1i}\dot{x}_{1i}+\ddot{x}_{2i}\dot{x}_{2i}+\ddot{x}_{3i}\dot{x}_{3i} - \phi^2(\dot{x}_{1i}x_{1i}+\dot{x}_{2i}x_{2i})] = 2\frac{dV}{dt} , \tag{4.1.7}$$

where the function V is given by (2.1.13). Since the potential V does not depend explicitly on time and it is only a function of coordinates x_{li} $(l=1,2,3;\ i=1(1)N)$, then integrating the equation (4.1.7), we get

$$\sum_{i=1}^{N} m_i [\sum_{l=1}^{3} \dot{x}_{li}^2 - \phi^2(x_{1i}^2+x_{2i}^2)] = 2V + 2C.$$

Hence

$$\frac{1}{2} \sum_{i=1}^{N} m_i [\sum_{l=1}^{3} v_{li}^2 - \phi^2(x_{1i}^2+x_{2i}^2)] - V = C. \tag{4.1.8}$$

But

$$E = \frac{1}{2} \sum_{i=1}^{N} m_i \sum_{l=1}^{3} v_{li}^2 - V$$

denotes the total energy of the system of N material points, and so the equality (4.1.8) can be rewritten as follows

$$E - \frac{\phi^2}{2} \sum_{i=1}^{N} m_i (x_{1i}^2+x_{2i}^2) = C. \tag{4.1.9}$$

The last equation is called the Jacobi integral and the constant C is said to be the Jacobi constant.

As has been mentioned earlier, the equations of motion in a rotating frame of reference have a particular weight in practice when N=3. One of the most important problems in this case is the three-body problem in which one of the masses is so small that it does not influence the motion of the remaining bodies, but is affected by them in the usual way. It is called a restricted three-body problem which, in the case when the motion of the body with infinitely small mass takes place in the plane of the motion of remaining bodies, it is said to be a restricted plane problem. The restricted three-body problem finds an application, among others, to the study of the motion of artificial or natural satellites in the gravitational field of a central planet.

Let us consider the following problem. Two material points with masses m_1 and m_2 revolve around the common center of mass and move along the circles according to their mutual gravitational attraction. We want to determine the motion of the third body, say P, assuming that the mass of P is such so small that it does not influence the motion of the remaining bodies. Moreover, we assume that the coordinates and components of velocity of P at the initial moment are known. This problem has beeen considered by Euler and later by Jacobi ([53]).

In order to solve the above problem let us put the origin of a rectangular coordinate system in the center of masses m_1 and m_2 (the equations of motion do not change!). Let us point the axis x_1 in such a way that it goes across both material points with masses m_1 and m_2 (see Fig.27). If the mean motion of these points, moving along the circles with radiuses ρ_1 and ρ_2 respectively, is equal to ν, then the frame $x_1x_2x_3$ is a rotating one with the constant angular velocity $\theta=\nu t$. This means that to describe the motion of P we can use the equations (4.1.5)-(4.1.6) in which, taking into account our assumptions, we have to substitute

$$x_{11} = -\frac{m_2}{m_1+m_2}(\rho_1+\rho_2), \quad x_{12} = \frac{m_1}{m_1+m_2}(\rho_1+\rho_2),$$
$$x_{21}=x_{31}=x_{22}=x_{32}=0,$$
$$r_{P1} = [(x_1-x_{11})^2+x_2^2+x_3^2]^{1/2}, \quad r_{P2} = [(x_1-x_{12})^2+x_2^2+x_3^2]^{1/2}.$$

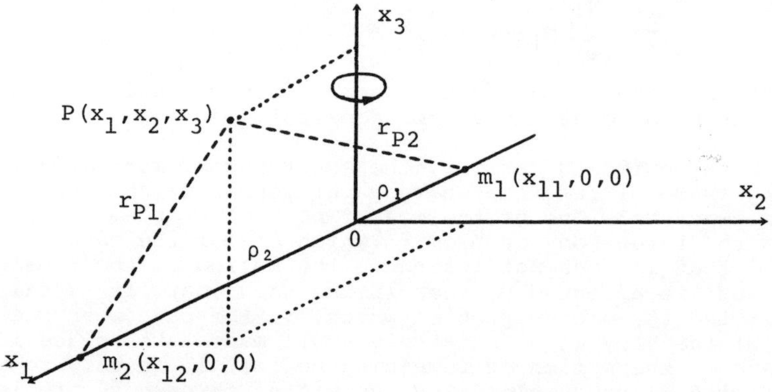

Fig.27. The motion of P in the frame of reference connected with m_1 and m_2

If we introduce the notations

$$\rho = \rho_1+\rho_2, \quad \mu = \frac{m_2}{m_1+m_2}, \tag{4.1.10}$$

then finally we get the following equations of the motion of P

$$\begin{aligned}
\ddot{x}_1 &= 2\nu\dot{x}_2+\nu^2x_1-G[m_1\frac{x_1+\mu\rho}{r_{P1}^3}+m_2\frac{x_1-(1-\mu)\rho}{r_{P2}^3}],\\
\ddot{x}_2 &= -2\nu\dot{x}_1+\nu^2x_2-G(m_1\frac{x_2}{r_{P1}^3}+m_2\frac{x_2}{r_{P2}^3}),\\
\ddot{x}_3 &= -G(m_1\frac{x_3}{r_{P1}^3}+m_2\frac{x_3}{r_{P2}^3}),
\end{aligned} \tag{4.1.11}$$

where

$$r_{P1} = [(x_1+\mu\rho)^2+x_2^2+x_3^2]^{1/2},$$

$$r_{P2} = \{[(x_1-(1-\mu)\rho]^2+x_2^2+x_3^2\}^{1/2}.$$

The mean motion ν can be determined from the third law of Kepler

$$\nu^2 = G\,\frac{m_1+m_2}{\rho^3}.$$

The Jacobi integral for the equations (4.1.11) has the form

$$\frac{1}{2}\sum_{l=1}^{3}\dot{x}_l^2 - G(\frac{m_1}{r_{P1}}+\frac{m_2}{r_{P2}}) - \frac{\nu^2}{2}(x_1^2+x_2^2) = C. \tag{4.1.12}$$

The equations (4.1.11) have been used to determine the trajectories of artificial satellites between the earth and the moon. They also have an application to the study of the motion of bodies with infinitely small masses nearby the so called equilibrium points. Let us take into account this problem.

If we take such units of distance and time that $\rho=1$ and $\nu=1$, then $Gm_1=1-\mu$ and $Gm_2=\mu$, where μ is given by (4.1.10). The Jacobi integral (4.1.12) can be written in the form

$$\frac{1}{2}(\dot{x}_1^2+\dot{x}_2^2+\dot{x}_3^2)-\frac{1-\mu}{r_{P1}}-\frac{\mu}{r_{P2}}-\frac{1}{2}(x_1^2+x_2^2) = C. \tag{4.1.13}$$

Since $V^2=\dot{x}_1^2+\dot{x}_2^2+\dot{x}_3^2$ is a square of the velocity of P, then from the above equation we have

$$2(\Omega+C) = V^2, \tag{4.1.14}$$

where $\Omega=(1-\mu)/r_{P1}+\mu/r_{P2}+(x_1^2+x_2^2)/2$.

The equation (4.1.14) for V=0 determines a plane on which the relative velocity is equal to zero. Since the function Ω is a sum of positive values r_{P1}^{-1}, r_{P2}^{-1} and $x_1^2+x_2^2$ multiplied by some positive coefficients, then $2(\Omega+C)\geq 0$. That means the body P can move only on one side of the plane $2(\Omega+C)=0$. Moreover, from (4.1.13) and (4.1.14) it follows that the planes of zero-velocities given by the equation

$$\frac{2(1-\mu)}{r_{P1}} + \frac{2\mu}{r_{P2}} + x_1^2 + x_2^2 + C = 0 \tag{4.1.15}$$

are symmetrical with respect to the planes x_1x_2 and x_1x_3, and if $m_1=m_2$ they are also symmetrical with respect to the plane x_2x_3.

On the plane defined by (4.1.15) there are singular points determined by the equations

$$\frac{\partial\Omega}{\partial x_l} = 0; \quad l=1,2,3. \tag{4.1.16}$$

Coordinates of these points have to fulfil the equations (4.1.15) and (4.1.16) at the same time. From the equations of the motion of P and from (4.1.16) it follows that in each singular point we have $\dot{x}_l=0$ and $\ddot{x}_l=0$ $(l=1,2,3)$. It means that a body with infinitely small mass, being in the singular point of a zero-velocity plane, has the velocity and acceleration equal to zero. Thus, these are the points of relative equilibrium of P with respect to the rotating frame of reference. In other words: a body with an infinitely small mass placed in an equilibrium point will be in this point forever.

From the equation $\partial\Omega/\partial x_3=0$ it follows that $x_3=0$. It means that the equilibrium points lie on the plane x_1x_2. Taking into account the condition $x_3=0$ in the equations (4.1.15) and (4.1.16), we get for the equilibrium points L_j $(j=4,5)$ the following values of coordinates $(a_{1j}, a_{2j}, 0)$ (for calculation details we refer e.g. to [22] or [91])

$$\begin{aligned} L_4:\quad & a_{14} = \frac{1}{2}\,\frac{m_1-m_2}{m_1+m_2}, \quad a_{24} = \frac{\sqrt{3}}{2}, \\ L_5:\quad & a_{15} = \frac{1}{2}\,\frac{m_1-m_2}{m_1+m_2}, \quad a_{25} = -\frac{\sqrt{3}}{2}. \end{aligned} \tag{4.1.17}$$

For the points L_j $(j=1,2,3)$ we have $a_{2j}=0$. In order to determine a_{1j} we have to solve the fifth order equation of the form

$$\alpha_0+\alpha_1 x+\alpha_2 x^2+\alpha_3 x^3+\alpha_4 x^4+\alpha_5 x^5 = 0, \tag{4.1.18}$$

where

$$\alpha_0 = \begin{cases} 2\mu^3-3\mu^2+3\mu-1, \text{ for } L_1, \\ -3\mu^2+3\mu-1, \text{ for } L_2, \\ 3\mu^2-3\mu+1, \text{ for } L_3, \end{cases}$$

$$\alpha_1 = \begin{cases} \mu^4-2\mu^3+5\mu^2-4\mu+2, \text{ for } L_1, \\ \mu^4-2\mu^3+\mu^2-4\mu+2, \text{ for } L_2, \\ \mu^4-2\mu^3+\mu^2+4\mu-2, \text{ for } L_3, \end{cases} \tag{4.1.19}$$

$$\alpha_2 = \begin{cases} 4\mu^3-6\mu^2+4\mu-1, \text{ for } L_1, \\ 4\mu^3-6\mu^2+2\mu-1, \text{ for } L_2, \\ 4\mu^3-6\mu^2+2\mu+1, \text{ for } L_3, \end{cases}$$

$$\alpha_3 = 6\mu^2-6\mu+1, \quad \alpha_4 = 4\mu-2, \quad \alpha_5 = 1.$$

The equation (4.1.18) has only one real root in each case. To solve this equation we can use the Newton iterative process. In this process it is convenient to take the value $1-\mu$ for the points L_1 and L_2, and the value $-1-\mu$ for the point L_3 as the initial approximation.

The positions of equilibrium points are presented in Fig.28.

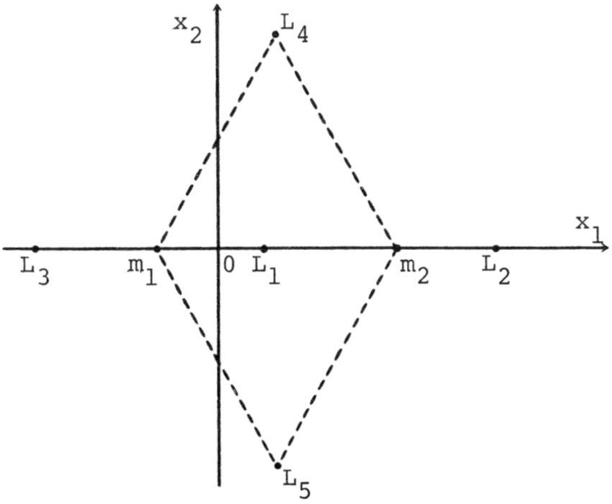

Fig.28. Positions of equilibrium points

The points L_1, L_2 and L_3 lie on the axis x_1 going across both masses m_1 and m_2. These points are called collinear equilibrium points. The points L_4 and L_5 are symmetrical with respect to the axis x_1 and they form equilateral triangles with the material points. These points are said to be triangular equilibrium points. In Table XXXV the coordinates of equilibrium points for the system Earth-Moon ($m_1=1$, $m_2=0.0123001$, $G=0.987849354$) and Sun-Jupiter ($m_1=1$, $m_2=0.0009546$ $G=0.999046316$) are given.

Table XXXV. Coordinates of equilibrium points

system	equilibrium point	a_{1j}	a_{2j}
Earth-Moon	L_1	0.836914829402	0.000000000000
	L_2	1.155682397100	0.000000000000
	L_3	-1.005062670905	0.000000000000
	L_4	0.487849354159	0.866025447845
	L_5	0.487849354159	-0.866025447845
Sun-Jupiter	L_1	0.932370120796	0.000000000000
	L_2	1.068825955574	0.000000000000
	L_3	-1.000397368491	0.000000000000
	L_4	0.499046315508	0.866025447845
	L_5	0.499046315508	-0.866025447845

In order to study the motion of a body with an infinitely small mass it is convenient to displace the origin of the rotating frame to the point L_j ($j=1(1)5$) with the coordinates $(a_{1j},a_{2j},0)$. Then from (4.1.11) we get

$$\begin{aligned}
\ddot{x}_1 &= 2\dot{x}_2+x_1+a_{1j}-G(m_1\frac{x_1+a_{1j}+\mu}{r_1^3}+m_2\frac{x_1+a_{1j}+\mu-1}{r_2^3}),\\
\ddot{x}_2 &= -2\dot{x}_1+x_2+a_{2j}-G(x_2+a_{2j})(\frac{m_1}{r_1^3}+\frac{m_2}{r_2^3}), \qquad (4.1.20)\\
\ddot{x}_3 &= -Gx_3(\frac{m_1}{r_1^3}+\frac{m_2}{r_2^3}),
\end{aligned}$$

where

$$\begin{aligned}
r_1 &= [(x_1+a_{1j}+\mu)^2+(x_2+a_{2j})^2+x_3^2]^{1/2},\\
r_2 &= [(x_1+a_{1j}+\mu-1)^2+(x_2+a_{2j})^2+x_3^2]^{1/2}.
\end{aligned}$$

The Jacobi integral for the above equations has the form

$$\frac{1}{2}\sum_{l=1}^{3}\dot{x}_l^2 - G\left(\frac{m_1}{r_1}+\frac{m_2}{r_2}\right) - \frac{1}{2}[(x_1+a_{1j})^2+(x_2+a_{2j})^2] = C,$$

where C=const. In (4.1.20) it is assumed that $\nu=1$ and the distance between the masses m_1 and m_2 is equal to $\rho=1$, so that $G=1/(m_1+m_2)$. Under some additional assumptions, from the equations (4.1.20) we can obtain periodical solutions (see e.g.[91]).

Another example of the application of (4.1.5) is the equation of motion of the moon given by Hill ([48]). If we place the origin of the frame in the center of the earth, assume that the plane x_1x_2 is in line with the sun's orbit plane and the frame rotates with the constant angular velocity ν', where ν' denotes the mean motion of the sun (see Fig.29), then the equations of motion of the moon have the form (see (4.1.5)-(4.1.6))

$$\begin{aligned} \ddot{x}_1-2\nu'\dot{x}_2-\nu'^2x_1 &= \frac{\partial U}{\partial x_1}, \\ \ddot{x}_2+2\nu'\dot{x}_1-\nu'^2x_2 &= \frac{\partial U}{\partial x_2}, \\ \ddot{x}_3 &= \frac{\partial U}{\partial x_3}, \end{aligned} \tag{4.1.21}$$

where

$$U = G\,\frac{m_o+m_1}{m_om_1}\left(\frac{m_om_1}{r_o}+\frac{m_1m_2}{r_1}+\frac{m_om_2}{r_2}\right),$$

and where the factor $(m_o+m_1)/(m_om_1)$ is related to the displacement of the frame's origin to the center of the earth. m_o denotes a mass of the earth, m_1 is a mass of the moon and m_2 denotes a mass of the sun. The function U can be presented as ([91])

$$U = G\,\frac{m_o+m_1}{r_o} + R, \tag{4.1.22}$$

where

$$R = \frac{Gm_2}{r'}\left(\frac{r_o}{r'}\right)^2\left(\frac{3}{2}\cos^2\gamma-\frac{1}{2}\right)+\ldots \tag{4.1.23}$$

is a perturbing function and where r' denotes the distance between the sun and the center of mass of the system earth-moon. By γ we denote in (4.1.23) the angle MBS (see Fig.29).

If we introduce a new function U' defined by

$$U' = U + \frac{1}{2}\nu'^2(x_1^2+x_2^2), \tag{4.1.24}$$

then the equations (4.1.21) have the form

$$\begin{aligned}\ddot{x}_1 - 2\nu'\dot{x}_2 &= \frac{\partial U'}{\partial x_1}, \\ \ddot{x}_2 + 2\nu'\dot{x}_1 &= \frac{\partial U'}{\partial x_2}, \\ \ddot{x}_3 &= \frac{\partial U'}{\partial x_3}.\end{aligned} \tag{4.1.25}$$

Introducing the parameters

$$M = \frac{\nu'}{\nu-\nu'}, \qquad \kappa = G\frac{m_o+m_1}{(\nu-\nu')^2}, \tag{4.1.26}$$

where ν is a mean motion of the moon, from (4.1.22)-(4.1.24) we obtain

$$\begin{aligned}U' &= (\nu-\nu')^2\Big[\frac{\kappa}{r_o} + \frac{Gm_2}{(\nu-\nu')^2 r'}\Big(\frac{r_o}{r'}\Big)^2\Big(\frac{3}{2}\cos^2\gamma - \frac{1}{2}\Big) \\ &\qquad + \frac{1}{2}M^2(x_1^2+x_2^2)+\ldots\Big] \\ &= (\nu-\nu')^2\Big[\frac{\kappa}{r_o} + \frac{M^2}{2}(3x_1^2-x_3^2) + \frac{Gm_2}{(\nu-\nu')^2 r'}\Big(\frac{r_o}{r'}\Big)^2\Big(\frac{3}{2}\cos^2\gamma - \frac{1}{2}\Big) \\ &\qquad - M^2\Big(x_1^2 - \frac{x_2^2}{2} - \frac{x_3^2}{2}\Big)+\ldots\Big].\end{aligned} \tag{4.1.27}$$

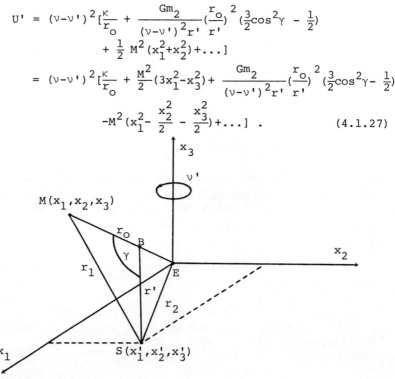

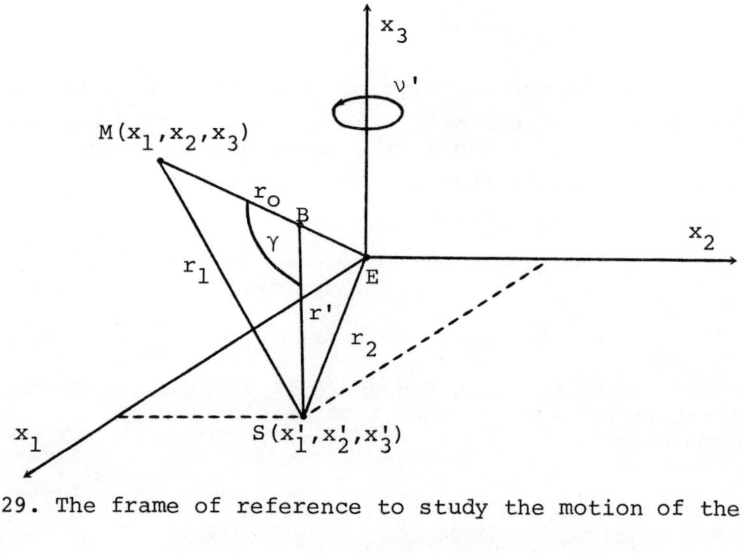

Fig.29. The frame of reference to study the motion of the moon

If we assume that in the considered frame of reference the axis x_1 goes across the center of the sun and that the sun revolves around the earth along the circular orbit, then

$$x_2' = 0, \quad x_1' = r', \quad r_o\cos\gamma = x_1, \quad r_2 = r'.$$

Moreover, from the Kepler law we have $Gm_2 = \nu'^2 r_2^3$ in this case. Thus

$$\frac{Gm_2}{(\nu-\nu')^2 r'}\left(\frac{r_o}{r'}\right)^2\left(\frac{3}{2}\cos^2\gamma - \frac{1}{2}\right) - M^2\left(x_1^2 - \frac{x_2^2}{2} - \frac{x_3^2}{2}\right) + \ldots$$

$$= M^2\left(\frac{r_2}{r'}\right)^3\left(\frac{3}{2}x_1^2 - \frac{1}{2}r_o^2\right) - M^2\left(x_1^2 - \frac{x_2^2}{2} - \frac{x_3^2}{2}\right) + \ldots$$

$$= M^2\left(\frac{3}{2}x_1^2 - \frac{x_1^2}{2} - \frac{x_2^2}{2} - \frac{x_3^2}{2}\right) - M^2\left(x_1^2 - \frac{x_2^2}{2} - \frac{x_3^2}{2}\right) + \ldots = 0.$$

Hence and from (4.1.27) it follows that

$$U' = (\nu-\nu')^2\left[\frac{\kappa}{r_o} + \frac{M^2}{2}(3x_1^2 - x_3^2)\right].$$

That means the equations of motion of the moon have the form (see (4.1.25))

$$\begin{aligned}
\ddot{x}_1 - 2\nu'\dot{x}_2 &= (\nu-\nu')^2\left(-\frac{\kappa x_1}{r^3} + 3M^2 x_1\right),\\
\ddot{x}_2 + 2\nu'\dot{x}_1 &= (\nu-\nu')^2\left(-\frac{\kappa x_2}{r^3}\right),\\
\ddot{x}_3 &= (\nu-\nu')^2\left(-\frac{\kappa x_3}{r^3} - M^2 x_3\right),
\end{aligned} \tag{4.1.28}$$

where $r \equiv r_3 = (x_1^2 + x_2^2 + x_3^2)^{1/2}$. If we introduce a new time unit τ defined by

$$\tau = (\nu-\nu')(t-t_o), \tag{4.1.29}$$

then the equations (4.1.28) can be written in the form

$$\begin{aligned}
&\frac{d^2x_1}{d\tau^2} - 2M\frac{dx_2}{d\tau} + \left(\frac{\kappa}{r^3} - 3M^2\right)x_1 = 0,\\
&\frac{d^2x_2}{d\tau^2} + 2M\frac{dx_1}{d\tau} + \frac{\kappa}{r^3}x_2 = 0,\\
&\frac{d^2x_3}{d\tau^2} + \left(\frac{\kappa}{r^3} + M^2\right)x_3 = 0.
\end{aligned} \tag{4.1.30}$$

The above equations are said to be the Hill equations. They have the following Jacobi integral

$$\sum_{l=1}^{3} \left(\frac{dx_l}{d\tau}\right)^2 - 2\,\frac{\kappa}{r} - M^2(3x_1^2-x_3^2) = C. \tag{4.1.31}$$

Solving (4.1.30), we obtain an inferior orbit of the moon which contains a considerable part of perturbations in comparison with the Keplerian orbit. Assuming also $x_3=0$ (we can do it, taking into account a small inclination of the moon's orbit to the ecliptic), we get a periodical solution which is called a variational curve (see e.g.[22] and [91]).

4.2. THE APPLICATION OF CONVENTIONAL NUMERICAL METHODS

The purpose of this section is to present some conventional algorithms used to solve the equations (4.1.11), (4.1.20) and (4.1.30). We will refer to the algorithms presented previously.

At first let us consider the equations (4.1.11). Introducing new unknown variables $v_l \equiv \dot{x}_l$ (l=1,2,3), we can rewrite these equations in the form

$$\dot{x}_l = v_l, \quad x_l(t_o) = x_l^o; \quad l=1,2,3; \tag{4.2.1}$$

$$\dot{v}_1 = 2\nu v_2+\nu^2 x_1-G[m_1\frac{x_1+\mu\rho}{r_{P1}^3} +m_2\frac{x_1-(1-\mu)\rho}{r_{P2}^3}], \quad v_1(t_o) = v_1^o,$$

$$\dot{v}_2 = -2\nu v_1+\nu^2 x_2-G(m_1\frac{x_2}{r_{P1}^3} +m_2\frac{x_2}{r_{P2}^3}), \quad v_2(t_o) = v_2^o,$$

$$\dot{v}_3 = -G(m_1\frac{x_3}{r_{P1}^3} +m_2\frac{x_3}{r_{P2}^3}), \quad v_3(t_o) = v_3^o,$$

where ν, ρ, r_{P1} and r_{P2} denote the values defined on page 171. The above equations are to be the system of six differential equations of the first order. In order to solve them numerically we ought to have the following input data: G, m_1, m_2, ρ, t_o, Δt (a basic step size of solution), n (the number of steps; $n=(T-t_o)/\Delta t$, where T denotes a final moment), x_l^o and v_l^o (l=1,2,3). In the algorithms presented below, x_l^k and

v_l^k $(l=1,2,3)$, i.e. the coordinates and components of velocity at the moments $t_k=t_o+k\Delta t$ $(k=1(1)n)$, are to be the results.

<u>Algorithm 19.</u> The modified Euler method with an automatic step size correction (compare Algorithm 1)

1° $\mu:=m_2/(m_1+m_2)$;

$\nu:=[G(m_1+m_2)/\rho^3]^{1/2}$;

2° $H:=\Delta t$;

$S:=$ <u>true</u>;

$k:=0$;

3° $t:=t_o+k\Delta t$;

4° for $l=1,2,3$:

$$x_l^{k+1}:=x_l^k+H\{v_l^k-\frac{HG}{2}[m_1\frac{x_l^k}{(r_{P1}^k)^3}+m_2\frac{x_l^k}{(r_{P2}^k)^3}]\},$$

where $r_{P1}^k=[(x_1^k+\mu\rho)^2+(x_2^k)^2+(x_3^k)^2]^{1/2}$,

$r_{P2}^k=\{[x_1^k+(\mu-1)\rho]^2+(x_2^k)^2+(x_3^k)^2\}^{1/2}$;

$$v_l^{k+1}:=v_l^k-HG[m_1\frac{x_l^k+\frac{H}{2}v_l^k}{(\tilde{r}_{P1}^k)^3}+m_2\frac{x_l^k+\frac{H}{2}v_l^k}{(\tilde{r}_{P2}^k)^3}],$$

where $\tilde{r}_{P1}^k=[(x_1^k+\frac{H}{2}v_1^k+\mu\rho)^2+(x_2^k+\frac{H}{2}v_2^k)^2+(x_3^k+\frac{H}{2}v_3^k)]^{1/2}$,

$\tilde{r}_{P2}^k=\{[x_1^k+\frac{H}{2}v_1^k+(\mu-1)\rho]^2+(x_2^k+\frac{H}{2}v_2^k)^2$
$+(x_3^k+\frac{H}{2}v_3^k)^3\}^{1/2}$;

$$x_1^{k+1}:=x_1^{k+1}+\frac{H^2}{2}\{2\nu v_2^k+\nu^2x_1^k-G[m_1\frac{\mu\rho}{(r_{P1}^k)^3}+m_2\frac{(\mu-1)\rho}{(r_{P2}^k)^3}]\};$$

$$x_2^{k+1}:=x_2^{k+1}+\frac{H^2}{2}(-2\nu v_1^k+\nu^2x_2^k);$$

$$v_1^{k+1}:=v_1^{k+1}+H[2\nu(v_2^k+\frac{H}{2}\{-2\nu v_1^k+\nu^2x_2^k-G[m_1\frac{x_2^k}{(r_{P1}^k)^3}+m_2\frac{x_2^k}{(r_{P2}^k)^3}]\})$$
$$+\nu^2(x_1^k+\frac{H}{2}v_1^k)-G[m_1\frac{\mu\rho}{(\tilde{r}_{P1}^k)^3}+m_2\frac{(\mu-1)\rho}{(\tilde{r}_{P2}^k)^3}]];$$

$$v_2^{k+1}:=v_2^{k+1}+H[-2\nu(v_1^k+\frac{H}{2}\{2\nu v_2^k+\nu^2x_1^k-G[m_1\frac{x_1^k+\mu\rho}{(r_{P1}^k)^3}+m_2\frac{x_1^k+(\mu-1)\rho}{(r_{P2}^k)^3}]\})$$
$$+\nu^2(x_2^k+\frac{H}{2}v_2^k)],$$

where r_{Ps}^k and $\tilde{r}_{Ps}^k$ (s=1 or s=2) are defined as above;

5° if S=<u>true</u> then:
 go to 6°;
 $H:=H_1$;
 S:=<u>true</u>;
 printout x_l^{k+1} and v_l^{k+1} (l=1,2,3);
 k:=k+1;
 if $k \le n-1$ then:
 go to 3°;
 stop;

6° F:=<u>true</u>;
 for l=1,2,3:
 $y_l:=x_l^k$;
 $w_l:=v_l^k$;

7° for l=1,2,3:

$$z_l:=y_l+\frac{H}{2}w_l-\frac{HG}{4}\left[m_1\frac{y_l}{r_{P1}^3}+m_2\frac{y_l}{r_{P2}^3}\right],$$

where $r_{P1}=[(y_1+\mu\rho)^2+y_2^2+y_3^2]^{1/2}$,

$r_{P2}=\{[y_1+(\mu-1)\rho]^2+y_2^2+y_3^2\}^{1/2}$;

$$u_l:=w_l-\frac{HG}{2}\left[m_1\frac{y_l+\frac{H}{4}w_l}{\tilde{r}_{P1}^3}+m_2\frac{y_l+\frac{H}{4}w_l}{\tilde{r}_{P2}^3}\right],$$

where $\tilde{r}_{P1}=[(y_1+\frac{H}{4}w_1+\mu\rho)^2+(y_2+\frac{H}{4}w_2)^2+(y_3+\frac{H}{4}w_3)^2]^{1/2}$,

$\tilde{r}_{P2}=\{[y_1+\frac{H}{4}w_1+(\mu-1)\rho]^2+(y_2+\frac{H}{4}w_2)^2$
$+(y_3+\frac{H}{4}w_3)^2\}^{1/2}$;

$$z_1:=z_1+\frac{H^2}{8}\{2\nu w_2+\nu^2 y_1-G[m_1\frac{\mu\rho}{r_{P1}^3}+m_2\frac{(\mu-1)\rho}{r_{P2}^3}]\};$$

$$z_2:=z_2+\frac{H^2}{8}(-2\nu w_1+\nu^2 y_2);$$

$$u_1:=u_1+\frac{H}{2}(2\nu\{w_2+\frac{H}{4}[-2\nu w_1+\nu^2 y_2-G(m_1\frac{y_2}{r_{P1}^3}+m_2\frac{y_2}{r_{P2}^3})]\}$$
$$+\nu^2(y_1+\frac{H}{4}w_1)-G[m_1\frac{\mu\rho}{r_{P1}^3}+m_2\frac{(\mu-1)\rho}{r_{P2}^3}]);$$

$$u_2:=u_2+\frac{H}{2}[-2\nu(w_1+\frac{H}{4}\{2\nu w_2+\nu^2 y_1-G[m_1\frac{y_1+\mu\rho}{r_{P1}^3}+m_2\frac{y_1+(\mu-1)\rho}{r_{P2}^3}]\})$$
$$+\nu^2(y_2+\frac{H}{4}w_2)],$$

where r_{P1}, r_{P2}, $\tilde{r}_{P1}$ and $\tilde{r}_{P2}$ are determined as above;

8° if F=false then:
go to 9°;
F:=false;
for l=1,2,3:
$y_l := z_l$;
$w_l := u_l$;
go to 7°;

9° $\tilde{h} := H/[\frac{4}{3} \max_{l=1,2,3} (|x_l^{k+1} - z_l|, |v_l^{k+1} - u_l|)/\varepsilon]^{1/3}$
(ε denotes a discretization error given beforehand);
if $\tilde{h} \leq H/3$ then:
$H := 2\tilde{h}$;
go to 4°;
if $\tilde{h} \geq H$ then:
if $t + 2H \leq t_o + (k+1)\Delta t$ then:
$H := 2H$;
go to 4°;
$H := t_o + (k+1)\Delta t - t$;
$H_1 := H$;
S:=false;
go to 4°;
for l=1,2,3:
$x_l^k := z_l$;
$v_l^k := u_l$;
$t := t + H$;
if $t + 2\tilde{h} < t_o + (k+1)\Delta t$ then:
$H := 2\tilde{h}$;
go to 4°;
$H := t_o + (k+1)\Delta t - t$;
$H_1 := 2\tilde{h}$;
S:=false;
go to 4°.

Algorithm 20. The Runge-Kutta method of fourth order with an automatic step size correction (compare Algorithm 2)

$1^\circ, 2^\circ, 3^\circ, 5^\circ, 6^\circ, 8^\circ$ - as above;
4° for l=1,2,3:
$K_l^1 := v_l^k$;
$K_{l+3}^1 := -Gx_l^k [\frac{m_1}{(r_{P1}^k)^3} + \frac{m_2}{(r_{P2}^k)^3}]$,

where $r_{P1}^{k}=[(x_1^k+\mu\rho)^2+(x_2^k)^2+(x_3^k)^2]^{1/2}$,

$r_{P2}^{k}=\{[x_1^k+(\mu-1)\rho]^2+(x_2^k)^2+(x_3^k)^2\}^{1/2}$;

$K_4^1:=K_4^1+2\nu v_2^k+\nu^2 x_1^k-G[m_1\frac{\mu\rho}{(r_{P1}^k)^3}+m_2\frac{(\mu-1)\rho}{(r_{P2}^k)^3}]$,

where r_{P1}^k and r_{P2}^k are determined as above;

$K_5^1:=K_5^1-2\nu v_2^k+\nu^2 x_2^k$;

for j=2,3:

for l=1,2,3:

$K_l^j:=v_l^k+\frac{H}{2}K_{l+3}^{j-1}$;

$K_{l+3}^j:=-G(x_l^k+\frac{H}{2}K_l^{j-1})[\frac{m_1}{(\bar{r}_{P1}^k)^3}+\frac{m_2}{(\bar{r}_{P2}^k)^3}]$,

where $\bar{r}_{P1}^k=[(x_1^k+\frac{H}{2}K_1^{j-1}+\mu\rho)^2+(x_2^k+\frac{H}{2}K_2^{j-1})^2+(x_3^k+\frac{H}{2}K_3^{j-1})^2]^{1/2}$,

$\bar{r}_{P2}^k=\{[x_1^k+\frac{H}{2}K_1^{j-1}+(\mu-1)\rho]^2+(x_2^k+\frac{H}{2}K_2^{j-1})^2+(x_3^k+\frac{H}{2}K_3^{j-1})^2\}^{1/2}$;

$K_4^j:=K_4^j+2\nu(v_2^k+\frac{H}{2}K_5^{j-1})+\nu^2(x_1^k+\frac{H}{2}K_1^{j-1})-G[m_1\frac{\mu\rho}{(\bar{r}_{P1}^k)^3}+m_2\frac{(\mu-1)\rho}{(\bar{r}_{P2}^k)^3}]$,

where $\bar{r}_{P1}^k$ and $\bar{r}_{P2}^k$ are determined as above;

$K_5^j:=K_5^j-2\nu(v_1^k+\frac{H}{2}K_4^{j-1})+\nu^2(x_2^k+\frac{H}{2}K_2^{j-1})$;

for l=1,2,3:

$K_l^4:=v_l^k+HK_{l+3}^3$;

$K_{l+3}^4:=-G(x_l^k+HK_l^3)[\frac{m_1}{(\tilde{r}_{P1}^k)^3}+\frac{m_2}{(\tilde{r}_{P2}^k)^3}]$,

where $\tilde{r}_{P1}^k=[(x_1^k+HK_1^3+\mu\rho)^2+(x_2^k+HK_2^3)^2+(x_3^k+HK_3^3)^2]^{1/2}$,

$\tilde{r}_{P2}^k=\{[x_1^k+HK_1^3+(\mu-1)\rho]^2+(x_2^k+HK_2^3)^2+(x_3^k+HK_3^3)^2\}^{1/2}$;

$K_4^4:=K_4^4+2\nu(v_2^k+HK_5^3)+\nu^2(x_1^k+HK_1^3)-G[m_1\frac{\mu\rho}{(\tilde{r}_{P1}^k)^3}+m_2\frac{(\mu-1)\rho}{(\tilde{r}_{P2}^k)^3}]$,

where $\tilde{r}_{P1}^k$ and $\tilde{r}_{P2}^k$ are given above;

$K_5^4:=K_5^4-2\nu(v_1^k+HK_4^3)+\nu^2(x_2^k+HK_2^3)$;

for $l=1,2,3$:

$$x_l^{k+1} := x_l^k + \frac{H}{6}(K_l^1 + 2K_l^2 + 2K_l^3 + K_l^4);$$

$$v_l^{k+1} := v_l^k + \frac{H}{6}(K_{l+3}^1 + 2K_{l+3}^2 + 2K_{l+3}^3 + K_{l+3}^4);$$

7° for $l=1,2,3$:

$$K_l^1 := w_l;$$

$$K_{l+3}^1 := -Gy_l\left(\frac{m_1}{r_{P1}^3} + \frac{m_2}{r_{P2}^3}\right),$$

where $r_{P1} = [(y_1+\mu\rho)^2 + y_2^2 + y_3^2]^{1/2}$,

$r_{P2} = \{[y_1+(\mu-1)\rho]^2 + y_2^2 + y_3^2\}^{1/2}$;

$$K_4^1 := K_4^1 + 2\nu w_2 + \nu^2 y_1 - G\left[m_1\frac{\mu\rho}{r_{P1}^3} + m_2\frac{(\mu-1)\rho}{r_{P2}^3}\right],$$

where r_{P1} and r_{P2} are given above;

$$K_5^1 := K_5^1 - 2\nu w_2 + \nu^2 y_2;$$

for $j=2,3$:

for $l=1,2,3$:

$$K_l^j := w_l + \frac{H}{2}K_{l+3}^{j-1};$$

$$K_{l+3}^j := -G\left(y_l + \frac{H}{2}K_l^{j-1}\right)\left(\frac{m_1}{\bar{r}_{P1}^3} + \frac{m_2}{\bar{r}_{P2}^3}\right),$$

where $\bar{r}_{P1} = [(y_1 + \frac{H}{2}K_1^{j-1} + \mu\rho)^2 + (y_2 + \frac{H}{2}K_2^{j-1})^2 + (y_3 + \frac{H}{2}K_3^{j-1})^2]^{1/2}$,

$\bar{r}_{P2} = \{[y_1 + \frac{H}{2}K_1^{j-1} + (\mu-1)\rho]^2 + (y_2 + \frac{H}{2}K_2^{j-1})^2 + (y_3 + \frac{H}{2}K_3^{j-1})^2\}^{1/2}$;

$$K_4^j := K_4^j + 2\nu\left(w_2 + \frac{H}{2}K_5^{j-1}\right) + \nu^2\left(y_1 + \frac{H}{2}K_1^{j-1}\right) - G\left[m_1\frac{\mu\rho}{\bar{r}_{P1}^3} + m_2\frac{(\mu-1)\rho}{\bar{r}_{P2}^3}\right],$$

where $\bar{r}_{P1}$ and $\bar{r}_{P2}$ are determined above;

$$K_5^j := K_5^j - 2\nu\left(w_1 + \frac{H}{2}K_4^{j-1}\right) + \nu^2\left(y_2 + \frac{H}{2}K_2^{j-1}\right);$$

for $l=1,2,3$:

$$K_l^4 := w_l + HK_{l+3}^3;$$

$$K_{l+3}^4 := -G(y_l + HK_l^3)\left(\frac{m_1}{\tilde{r}_{P1}^3} + \frac{m_2}{\tilde{r}_{P2}^3}\right),$$

where $\tilde{r}_{P1} = [(y_1 + HK_1^3 + \mu\rho)^2 + (y_2 + HK_2^3)^2 + (y_3 + HK_3^3)^2]^{1/2}$,

$\tilde{r}_{P2} = \{[y_1 + HK_1^3 + (\mu-1)\rho]^2 + (y_2 + HK_2^3)^2 + (y_3 + HK_3^3)\}^{1/2}$;

$K_4^4:=K_4^4+2\nu(w_2+HK_5^3)+\nu^2(y_1+HK_1^3)-G[m_1\frac{\mu\rho}{\tilde{r}_{P1}^3}+m_2\frac{(\mu-1)\rho}{\tilde{r}_{P2}^3}]$,

where $\tilde{r}_{P1}$ and $\tilde{r}_{P2}$ are given above;

$K_5^4:=K_5^4+2\nu(w_1+HK_4^3)+\nu^2(y_2+HK_2^3)$;

for l=1,2,3:

$z_l:=y_l+\frac{H}{6}(K_l^1+2K_l^2+2K_l^3+K_l^4)$;

$u_l:=w_l+\frac{H}{6}(K_{l+3}^1+2K_{l+3}^2+2K_{l+3}^3+K_{l+3}^4)$;

9° $\tilde{h}:=H/[\frac{16}{15}\max_{l=1,2,3}(|x_l^{k+1}-z_l|,|v_l^{k+1}-u_l|)/\varepsilon]^{1/5}$;

further as in Algorithm 19, 9°.

<u>Algorithm 21.</u> The Butcher method of sixth order

1° compute x_l^1 and v_l^1 (l=1,2,3) by the Runge-Kutta method of fourth order with an automatic step size correction (see Algorithm 20);

2° k:=1;

3° for l=1,2,3:

$z_l:=x_l^{k-1}+\frac{\Delta t}{8}(9v_l^k+3v_l^{k-1})$;

$$w_l:=v_l^{k-1}-\frac{\Delta t\times G}{8}m_1[\frac{9x_l^k}{(r_{P1}^k)^3}+\frac{3x_l^{k-1}}{(r_{P1}^{k-1})^3}]+m_2[\frac{9x_l^k}{(r_{P2}^k)^3}+\frac{3x_l^{k-1}}{(r_{P2}^{k-1})^3}];$$

$$w_1:=w_1+\frac{\Delta t}{8}(2\nu(9v_2^k+3v_2^{k-1})+\nu^2(9x_1^k+3x_1^{k-1})$$
$$-G\{m_1\mu\rho[\frac{9}{(r_{P1}^k)^3}+\frac{3}{(r_{P1}^{k-1})^3}]+m_2(\mu-1)\rho[\frac{9}{(r_{P2}^k)^3}+\frac{3}{(r_{P2}^{k-1})^3}]\}),$$

where $r_{P1}^s=[(x_1^s+\mu\rho)^2+(x_2^s)^2+(x_3^s)^2]^{1/2}$,

$r_{P2}^s=\{[x_1^s+(\mu-1)\rho]^2+(x_2^s)^2+(x_3^s)^2\}^{1/2}$ (s=k or s=k-1);

$w_2:=w_2+\frac{\Delta t}{8}[-2\nu(9v_1^k+3v_1^{k-1})+\nu^2(9x_2^k+3x_2^{k-1})]$;

4° for l=1,2,3:

$\tilde{z}_l:=5.6x_l^k-4.6x_l^{k-1}+\frac{\Delta t}{1.5}(3.2w_l-6.0v_l^k-2.6v_l^{k-1})$;

$$f_l:=-G\{3.2z_l(\frac{m_1}{\tilde{r}_{P1}^3}+\frac{m_2}{\tilde{r}_{P2}^3})-6.0x_l^k[\frac{m_1}{(r_{P1}^k)^3}+\frac{m_2}{(r_{P2}^k)^3}]$$
$$-2.6x_l^{k-1}[\frac{m_1}{(r_{P1}^{k-1})^3}+\frac{m_2}{(r_{P2}^{k-1})^3}]\},$$

where $\tilde{r}_{P1}=[(z_1+\mu\rho)^2+z_2^2+z_3^2]^{1/2}$,

$\tilde{r}_{P2}=\{[z_1+(\mu-1)\rho]^2+z_2^2+z_3^2\}^{1/2}$,

$r_{P1}^s=[(x_1^s+\mu\rho)^2+(x_2^s)^2+(x_3^s)^2]^{1/2}$,

$r_{P2}^s=\{[x_1^s+(\mu-1)\rho]^2+(x_2^s)^2+(x_3^s)^2\}^{1/2}$ (s=k or s=k-1);

$$f_1:=f_1+2\nu(3.2w_2-6.0v_2^k-2.6v_2^{k-1})+\nu^2(3.2z_1-6.0x_1^k-2.6x_1^{k-1})$$
$$-G\{m_1\mu\rho[\frac{3.2}{\tilde{r}_{P1}^3}-\frac{6.0}{(r_{P1}^k)^3}-\frac{2.6}{(r_{P1}^{k-1})^3}]$$
$$m_2(\mu-1)\rho[\frac{3.2}{\tilde{r}_{P2}^3}-\frac{6.0}{(r_{P2}^k)^3}-\frac{2.6}{(r_{P2}^{k-1})^3}]\},$$

where $\tilde{r}_{P1}$, $\tilde{r}_{P2}$, r_{P1}^s and r_{P2}^s are determined as previously;

$$f_2:=f_2-2\nu(3.2w_1-6.0v_1^k-2.6v_1^{k-1})+\nu^2(3.2z_2-6.0x_2^k-2.6x_2^{k-1});$$

for l=1,2,3:

$$\tilde{w}_l:=5.6v_l^k-4.6v_l^{k-1}+\frac{\Delta t}{1.5}f_l;$$

5° for l=1,2,3:

$$x_l^{k+1}:=\frac{1}{3.1}(3.2x_l^k-0.1x_l^{k-1})+\frac{\Delta t}{9.3}(6.4w_l+9.6v_l^k-7.0v_l^{k-1}+\Delta t f_l);$$

$$v_l^{k+1}:=\frac{1}{3.1}(3.2v_l^k-0.1v_l^{k-1})$$
$$-\frac{\Delta t\times G}{9.3}\{6.4z_l(\frac{m_1}{\tilde{r}_{P1}^3}+\frac{m_2}{\tilde{r}_{P2}^3})+1.2x_l^k[\frac{m_1}{(r_{P1}^k)^3}+\frac{m_2}{(r_{P2}^k)^3}]$$
$$-0.1x_l^{k-1}[\frac{m_1}{(r_{P1}^{k-1})^3}+\frac{m_2}{(r_{P2}^{k-1})^3}]$$
$$+1.5\tilde{z}_l(\frac{m_1}{\tilde{\tilde{r}}_{P1}^3}+\frac{m_2}{\tilde{\tilde{r}}_{P2}^3})\},$$

where $\tilde{r}_{P1}$, $\tilde{r}_{P2}$, r_{P1}^s and r_{P2}^s are determined as above and

$\tilde{\tilde{r}}_{P1}=[(\tilde{z}_1+\mu\rho)^2+\tilde{z}_2^2+\tilde{z}_3^2]^{1/2}$,

$\tilde{\tilde{r}}_{P2}=\{[\tilde{z}_1+(\mu-1)\rho]^2+\tilde{z}_2^2+\tilde{z}_3^2\}^{1/2}$;

$$v_1^{k+1}:=v_1^{k+1}+\frac{\Delta t}{9.3}(2\nu(6.4w_2+1.2v_2^k-0.1v_2^{k-1}+1.5\tilde{w}_2)$$
$$+\nu^2(6.4z_1+1.2x_1^k-0.1x_1^{k-1}+1.5\tilde{z}_1)$$
$$-G\{m_1\mu\rho[\frac{6.4}{\tilde{r}_{P1}^3}+\frac{1.2}{(r_{P1}^k)^3}-\frac{0.1}{(r_{P1}^{k-1})^3}+\frac{1.5}{\tilde{\tilde{r}}_{P1}^3}]$$

$$+m_2(\mu-1)\rho[\frac{6.4}{r_{P2}^3}+\frac{1.2}{(r_{P2}^k)^3}-\frac{0.1}{(r_{P2}^{k-1})^3}+\frac{1.5}{\tilde{r}_{P2}^3}]),$$

where $\tilde{r}_{P1}$, $\tilde{r}_{P2}$, r_{P1}^s, r_{P2}^s, $\tilde{\tilde{r}}_{P1}$ and $\tilde{\tilde{r}}_{P2}$ are given above;

$$v_2^{k+1}:=v_2^{k+1}+\frac{\Delta t}{9.3}[-2\nu(6.4w_1+1.2v_1^k-0.1v_1^{k-1}+1.5\tilde{w}_1)$$
$$+\nu^2(6.4z_2+1.2x_2^k-0.1x_2^{k-1}+1.5\tilde{z}_2)];$$

6° printout x_l^{k+1} and v_l^{k+1} $(l=1,2,3)$;
k:=k+1;
if k<n then go to 3°;
stop.

In order to solve numerically the equations of motion of a body with an infinitely small mass near the equilibrium points, we write these equations in the form

$$\dot{x}_l = v_l, \quad x_l(t_o) = x_l^o; \quad l=1,2,3;$$
$$\dot{v}_1 = 2v_2+x_1+a_{1j}-G(m_1\frac{x_1+a_{1j}+\mu}{r_1^3}+m_2\frac{x_1+a_{1j}+\mu-1}{r_2^3}),$$
$$v_1(t_o) = v_1^o, \tag{4.2.2}$$
$$\dot{v}_2 = -2v_1+x_2+a_{2j}-G(x_2+a_{2j})(\frac{m_1}{r_1^3}+\frac{m_2}{r_2^3}), \quad v_2(t_o) = v_2^o,$$
$$\dot{v}_3 = -Gx_3(\frac{m_1}{r_1^3}+\frac{m_2}{r_2^3}), \quad v_3(t_o) = v_3^o,$$

where the values μ, r_1, r_2, a_{1j} and a_{2j} $(j=1(1)5)$ are defined as previously (see pages 171-174). Let us remember that the values a_{1j} $(j=1,2,3)$ are computed using the Newton iteration process. Since the equations (4.2.2) do not differ much from the equations (4.2.1), we leave the composition of algorithms in the reader's hands.

To find an algorithm for the Hill equations (4.1.30), it is convenient to write them as follows

$$\dot{x}_l = v_l, \quad x_l(t_o) = x_l^o; \quad l=1,2,3;$$
$$\dot{v}_1 = 2Mv_2+(3M^2-\varkappa/r^3)x_1, \quad v_1(t_o) = v_1^o, \tag{4.2.3}$$
$$\dot{v}_2 = -2Mv_1-\varkappa x_2/r^3, \quad v_2(t_o) = v_2^o,$$
$$\dot{v}_3 = -(M^2+\varkappa/r^3)x_3, \quad v_3(t_o) = v_3^o,$$

where $r=(\sum_{l=1}^{3} x_l^2)^{1/2}$. The constants M and $\varkappa$ are determined by (4.1.26). It should be noted that in the above equations the dot denotes a differentiation with respect to the time τ given by (4.1.29). The composition of algorithms in this case we also leave in the reader's hands.

The application of the modified Euler method with an automatic step size correction and the Butcher method of sixth order to the equations (4.2.2) and (4.2.3) yields the results presented in Tables XXXVI-XXXIX. In the case of equations (4.2.2) we have taken the equilibrium points of the earth--moon system, have assumed the mean distance between the earth and the moon (i.e. 384000 km) as a unit of distance and have taken the mass of the earth as a unit of mass. Then $m_2=0.0123001$ and $G=1/(m_1+m_2)=0.987849354$. For Hill's equations it was assumed that $\nu=2.661699489\times10^{-6}$, $\nu'=1.99098870\times10^{-7}$, $M=0.080848941681$, $\varkappa=1.171417824464$.

Table XXXVI. The solution of (4.2.2) by the modified Euler method with an automatic step size correction ($\Delta t=0.01$, $\varepsilon=0.0000000001$)

point	k	l	$\dot{x}_l\times10^4$	$v_l\times10^4$	Jacobi's constant characteristics
L_1	0	1	1.00000000	0.15000000	C=-1.594170609387
		2	1.00000000	-7.25000000	
	115	1	-0.56747023	-1.23330580	α=8 , β=3
		2	-2.19051187	5.26032987	C=-1.594170609387
	230	1	2.22291215	7.53270848	α=8 , β=3
		2	2.07177820	-5.16665413	C=-1.594170609386
					CPU time 0.50 sec
L_4	0	1	1.00000000	2.20000000	C=-1.493998489152
		2	1.00000000	-1.45000000	
	150	1	2.48409394	0.61180397	α=8 , β=3
		2	-0.64114283	-1.59653793	C=-1.493998489152
	300	1	1.80738648	-1.29487070	α=8 , β=3
		2	-1.51474232	-0.63178897	C=-1.493998489152
					CPU time 0.66 sec

Notations: k - moment, l - number of axis, C - Jacobi's constant, α - number of calculations of right-hand sides of (4.2.2) from the moment k-1 to the moment k, β - number of integration steps.

Table XXXVII. The solution of (4.2.2) by the Butcher method of sixth order (the data at the initial moments as in Table XXXVI)

point	k	l	$x_l \times 10^4$	$v_l \times 10^4$	Jacobi's constant characteristics
L_1	115	1	-0.56743311	-1.23335893	$\alpha=4$, $\beta=1$
		2	-2.19063038	5.26006400	C=-1.594170609387
	230	1	2.22323678	7.53399644	$\alpha=4$, $\beta=1$
		2	2.07170346	-5.16647184	C=-1.594170609387
					CPU time 0.43 sec
L_4	150	1	2.48386000	0.61252911	$\alpha=4$, $\beta=1$
		2	-0.64059025	-1.59669310	C=-1.493998489152
	300	1	1.80825366	-1.29359457	$\alpha=4$, $\beta=1$
		2	-1.51428298	-0.63280039	C=-1.493998489152
					CPU time 0.58 sec

Table XXXVIII. The solution of Hill's equations by the modified Euler method with an automatic step size correction ($\Delta\tau$=0.0531922, ε=0.00000001)

k	l	x_l	v_l	Jacobi's constant characteristics
0	1	-0.9447782000	0.3286969000	C=-1.3836049279
	2	-0.2673999000	-0.9500594000	
	3	-0.0859043000	0.0127522000	
100	1	-0.9059322350	-0.3408047328	$\alpha=194$, $\beta=65$
	2	0.3989881311	-0.9347395580	C=-1.3838999375
	3	-0.0862877753	-0.0074280426	
200	1	-0.4482708739	-0.8374808920	$\alpha=164$, $\beta=55$
	2	0.9009837880	-0.4872502574	C=-1.3841972077
	3	-0.0844465141	-0.0288690524	

CPU time 4.10 sec

Table XXXIX. The solution of Hill's equations by the Butcher method of sixth order (the data at the initial moment as in Table XXXVIII)

k	l	x_l	v_l	Jacobi's constant characteristics
100	1	-0.9055669211	-0.3422732257	$\alpha=4$, $\beta=1$
	2	0.4006821250	-0.9339211910	C=-1.3835816463
	3	-0.0862987580	-0.0075841081	
200	1	-0.4439390539	-0.8394461614	$\alpha=4$, $\beta=1$
	2	0.9040925577	-0.4823523170	C=-1.3835814843
	3	-0.0843516764	-0.0293423259	

CPU time 0.35 sec

4.3. DISCRETE MECHANICS IN A ROTATING FRAME

Let us take into account the rectangular frame of reference rotating around the axis x_3 with a constant angular velocity ϕ. The motion of N material points in such a frame is described by (4.1.5)-(4.1.6). Replacing the derivatives in these equations by the differential quotients, we obtain

$$
\begin{aligned}
\frac{x_{1i}^{k+1}-x_{1i}^{k}}{\Delta t} - 2\phi\frac{x_{2i}^{k+1}-x_{2i}^{k}}{\Delta t} - \phi^2 x_{1i}^{k} &= F_{1i}(x_1^k, x_2^k, x_3^k)/m_i, \\
\frac{x_{2i}^{k+1}-x_{2i}^{k}}{\Delta t} + 2\phi\frac{x_{1i}^{k+1}-x_{1i}^{k}}{\Delta t} - \phi^2 x_{2i}^{k} &= F_{2i}(x_1^k, x_2^k, x_3^k)/m_i, \\
\frac{x_{3i}^{k+1}-x_{3i}^{k}}{\Delta t} &= F_{3i}(x_1^k, x_2^k, x_3^k)/m_i; \\
k=0,1,2,\ldots;\ i=1(1)N;
\end{aligned} \tag{4.3.1}
$$

where $\Delta t = t_{k+1}-t_k$ denotes the sufficiently small time interval and F_{1i} is given by (4.1.6). If the terms convergent to zero as $\Delta t \to 0$ are left out of account, then we may write

$$
\frac{x_{1i}^{k+1}-x_{1i}^{k}}{\Delta t} = v_{1i}^{k}, \quad x_{1i}^{k} = \frac{x_{1i}^{k+1}+x_{1i}^{k}}{2}, \quad v_{1i}^{k} = \frac{v_{1i}^{k+1}+v_{1i}^{k}}{2}.
$$

Moreover, applying to F_{1i} a discretization described in Sect.2.3.1, from (4.3.1) we get

$$
\begin{aligned}
\frac{x_{1i}^{k+1}-x_{1i}^{k}}{\Delta t} &= \frac{v_{1i}^{k+1}+v_{1i}^{k}}{2}; \quad 1=1,2,3; \\
\frac{v_{1i}^{k+1}-v_{1i}^{k}}{\Delta t} &= \phi(v_{2i}^{k+1}+v_{2i}^{k}) + \frac{\phi^2}{2}(x_{1i}^{k+1}+x_{1i}^{k}) + \frac{F_{1i}^{k}}{m_i}, \\
\frac{v_{2i}^{k+1}-v_{2i}^{k}}{\Delta t} &= -\phi(v_{1i}^{k+1}+v_{1i}^{k}) + \frac{\phi^2}{2}(x_{2i}^{k+1}+x_{2i}^{k}) + \frac{F_{2i}^{k}}{m_i}, \\
\frac{v_{3i}^{k+1}-v_{3i}^{k}}{\Delta t} &= \frac{F_{3i}^{k}}{m_i}; \quad i=1(1)N; \quad k=0,1,2,\ldots;
\end{aligned} \tag{4.3.2}
$$

where F_{1i}^{k} is determined by (2.3.1). The above equations are discrete equations of motion of N material points in a rotating frame of reference.

Let us now show that the equations (4.3.3) conserve the Jacobi constant of motion.

Theorem 4.1. If the motion of N material points in the frame $x_1x_2x_3$ rotating around x_3-axis is described by the equations (4.3.2), then

$$T^k - V^k - \frac{\phi^2}{2} \sum_{i=1}^{N} m_i[(x_{1i}^k)^2 + (x_{2i}^k)^2] = C \qquad (4.3.3)$$

for each k=0,1,2,..., where

$$T^k = \frac{1}{2} \sum_{i=1}^{N} m_i \sum_{l=1}^{3} (v_{li}^k)^2,$$
$$V^k = \frac{1}{2} \sum_{i=1}^{N} \sum_{\substack{j=1 \\ j\neq i}}^{N} \frac{m_i m_j}{[\sum_{l=1}^{3} (x_{li}^k - x_{lj}^k)^2]^{1/2}}. \qquad (4.3.4)$$

Proof. In order to prove (4.3.3) it is sufficient to show that for each k,

$$T^{k+1} - T^k = V^{k+1} - V^k + \frac{\phi^2}{2} \sum_{i=1}^{N} m_i[(x_{1i}^{k+1})^2 - (x_{1i}^k)^2 + (x_{2i}^{k+1})^2 - (x_{2i}^k)^2].$$

From (4.3.2) and (4.3.4) we have

$$T^{k+1} - T^k = \frac{1}{2} \sum_{i=1}^{N} m_i \sum_{l=1}^{3} [(v_{li}^{k+1})^2 - (v_{li}^k)^2]$$
$$= \frac{1}{2} \sum_{i=1}^{N} m_i \sum_{l=1}^{3} (v_{li}^{k+1} - v_{li}^k)(v_{li}^{k+1} + v_{li}^k)$$
$$= \sum_{i=1}^{N} m_i \{[\phi(v_{2i}^{k+1} + v_{2i}^k) + \frac{\phi^2}{2}(x_{1i}^{k+1} + x_{1i}^k) + \frac{F_{1i}^k}{m_i}](x_{1i}^{k+1} - x_{1i}^k)$$
$$+ [-\phi(v_{1i}^{k+1} + v_{1i}^k) + \frac{\phi^2}{2}(x_{2i}^{k+1} + x_{2i}^k) + \frac{F_{2i}^k}{m_i}](x_{2i}^{k+1} - x_{2i}^k)$$
$$+ \frac{F_{3i}^k}{m_i}(x_{3i}^{k+1} - x_{3i}^k)\}$$
$$= \sum_{i=1}^{N} F_i^k \circ (x_i^{k+1} - x_i^k)$$
$$+ \frac{\phi^2}{2} \sum_{i=1}^{N} m_i [(x_{1i}^{k+1} + x_{1i}^k)(x_{1i}^{k+1} - x_{1i}^k) + (x_{2i}^{k+1} + x_{2i}^k)(x_{2i}^{k+1} - x_{2i}^k)],$$

where ° denotes a scalar product. In Th.2.3 we have proved that

$$\sum_{i=1}^{N} F_i^k \circ (x_i^{k+1} - x_i^k) = V^{k+1} - V^k.$$

It means that the proof of (4.3.3) is finished.

For the discrete mechanics formulas in a rotating frame an analogous analysis of consistency, stability and convergence can be carried out as in Sect.2.3.2. To do it let us assume, without loss of generality, that the method (4.3.2) is applied to the solution of a problem on the time interval [0,1] and $\Delta t=1/n$. Using the notations from Sect.1.2, the method can be written in the form

$$\Phi_n \eta\left(\frac{\nu}{n}\right) = \begin{cases} \eta_p(0)-z_p^o, \quad \nu=0, \\ \eta_{p+3N}(0)-z_{p+3N}^o, \quad \nu=0, \\ \dfrac{\eta_p(\frac{\nu}{n})-\eta_p(\frac{\nu-1}{n})}{1/n} - [\eta_{p+3N}(\frac{\nu}{n})+\eta_{p+3N}(\frac{\nu-1}{n})]/2, \quad \nu=1(1)n, \\ \dfrac{\eta_{q+3N}(\frac{\nu}{n})-\eta_{q+3N}(\frac{\nu-1}{n})}{1/n} - \phi\,[\eta_{q+1+3N}(\frac{\nu}{n})+\eta_{q+1+3N}(\frac{\nu-1}{n})] \\ \qquad - \dfrac{\phi^2}{2}[\eta_q(\frac{\nu}{n})+\eta_q(\frac{\nu-1}{n})]-F_q[\eta(\frac{\nu}{n}),\eta(\frac{\nu-1}{n})], \\ \qquad \nu=1(1)n, \\ \dfrac{\eta_{q+1+3N}(\frac{\nu}{n})-\eta_{q+1+3N}(\frac{\nu-1}{n})}{1/n} + \phi\,[\eta_{q+3N}(\frac{\nu}{n})+\eta_{q+3N}(\frac{\nu-1}{n})] \\ \qquad + \dfrac{\phi^2}{2}[\eta_{q+1}(\frac{\nu}{n})+\eta_{q+1}(\frac{\nu-1}{n})]-F_{q+1}[\eta(\frac{\nu}{n}),\eta(\frac{\nu-1}{n})], \\ \qquad \nu=1(1)n, \\ \dfrac{\eta_{q+2+3N}(\frac{\nu}{n})-\eta_{q+2+3N}(\frac{\nu-1}{n})}{1/n} - F_{q+2}[\eta(\frac{\nu}{n}),\eta(\frac{\nu-1}{n})], \\ \qquad \nu=1(1)n; \end{cases} \tag{4.3.5}$$

$$p=1(1)3N; \quad q=1(3)3N-2;$$

where $F_q \equiv F^k_{li}/m_i$ and where F^k_{li} is given by (2.3.1). The initial value problem corresponding to the discrete problem (4.3.5) is of the form (compare (4.3.2))

$$\Phi y = \begin{cases} y_p(0)-z_p^o, \\ y_{p+3N}(0)-z_{p+3N}^o, \\ y'_p-y_{p+3N}, \\ y'_{q+3N}-2\phi y_{q+1+3N}-\phi^2 y_q-f_q(y), \\ y'_{q+1+3N}+2\phi y_{q+3N}-\phi^2 y_{q+1}-f_{q+1}(y), \\ y'_{q+2+3N}-f_{q+2}(y); \end{cases} \tag{4.3.6}$$

$$p=1(1)3N; \quad q=1(1)3N-2;$$

where $f_q \equiv F_{1i}/m_i$ and where F_{1i} is determined by (4.1.5).

It is easy to prove the following theorem (see the proofs of Th.2.4 and Th.2.5):

Theorem 4.2. The method (4.3.5) of discrete mechanics in a rotating frame of reference is consistent with the initial value problem (4.3.6). Moreover, if there exist the constants (2.3.23), then this method is of the first order.

Moreover, we have:

Theorem 4.3. If there exist the constants (2.3.23), then for $n>\phi+\phi^2/2+3N\tilde{M}$, where $\tilde{M}$ denotes a constant occuring in Lemma 2.2 and $\phi\geq 0$ is an angular velocity of rotation of the frame with respect to the axis x_3, the discrete problem (4.3.5) is stable on the initial value problem (4.3.6) with

$$S = e^{\alpha}\left(\frac{[\alpha]-\alpha+1}{[\alpha]+1}\right)^{-[\alpha]-1},$$

where S is a constant occuring in Def.1.5, $\alpha=\phi+\phi^2/2+3N\tilde{M}$ and [.] denotes an entire part.

Proof. Let $\varepsilon_p(\frac{\nu}{n})$ and $\delta_p(\frac{\nu}{n})$ $(p=1(1)3N)$ denote the quantities given by (2.3.29). We have

$$\varepsilon_p(\tfrac{\nu}{n}) = \varepsilon_p(\tfrac{\nu-1}{n}) + \frac{1}{2n}[\varepsilon_{p+3N}(\tfrac{\nu}{n})+\varepsilon_{p+3N}(\tfrac{\nu-1}{n})] + \frac{1}{n}\delta_p(\tfrac{\nu}{n}),$$

$$\begin{aligned}\varepsilon_{q+3N}(\tfrac{\nu}{n}) = {} & \varepsilon_{q+3N}(\tfrac{\nu-1}{n}) + \frac{\phi}{n}[\varepsilon_{q+1+3N}(\tfrac{\nu}{n})+\varepsilon_{q+1+3N}(\tfrac{\nu-1}{n})] \\ & + \frac{\phi^2}{2n}[\varepsilon_q(\tfrac{\nu}{n})+\varepsilon_q(\tfrac{\nu-1}{n})] + \frac{1}{n}\{F_q[\eta^{(1)}(\tfrac{\nu}{n}),\eta^{(1)}(\tfrac{\nu-1}{n})] \\ & -F_q[\eta^{(2)}(\tfrac{\nu}{n}),\eta^{(2)}(\tfrac{\nu-1}{n})]\} \\ & + \frac{1}{n}\delta_{q+3N}(\tfrac{\nu}{n}),\end{aligned}$$

$$\begin{aligned}\varepsilon_{q+1+3N}(\tfrac{\nu}{n}) = {} & \varepsilon_{q+1+3N}(\tfrac{\nu-1}{n}) - \frac{\phi}{n}[\varepsilon_{q+3N}(\tfrac{\nu}{n})+\varepsilon_{q+3N}(\tfrac{\nu-1}{n})] \\ & + \frac{\phi^2}{2n}[\varepsilon_{q+1}(\tfrac{\nu}{n})+\varepsilon_{q+1}(\tfrac{\nu-1}{n})] + \frac{1}{n}\{F_{q+1}[\eta^{(1)}(\tfrac{\nu}{n}),\eta^{(1)}(\tfrac{\nu-1}{n})] \\ & -F_{q+1}[\eta^{(2)}(\tfrac{\nu}{n}),\eta^{(2)}(\tfrac{\nu-1}{n})]\} \\ & + \frac{1}{n}\delta_{q+1+3N}(\tfrac{\nu}{n}),\end{aligned}$$

$$\begin{aligned}\varepsilon_{q+2+3N}(\tfrac{\nu}{n}) = {} & \varepsilon_{q+2+3N}(\tfrac{\nu-1}{n}) + \frac{1}{n}\{F_{q+2}[\eta^{(1)}(\tfrac{\nu}{n}),\eta^{(1)}(\tfrac{\nu-1}{n})] \\ & -F_{q+2}[\eta^{(2)}(\tfrac{\nu}{n}),\eta^{(2)}(\tfrac{\nu-1}{n})]\}\end{aligned}$$

$$+ \frac{1}{n} \delta_{q+2+3N}(\tfrac{\nu}{n});$$

$$p=1(1)3N;\ q=1(3)3N-2.$$

From the above equations and from Lemma 2.2 we obtain

$$|\varepsilon_p(\tfrac{\nu}{n})| \le |\varepsilon_p(\tfrac{\nu-1}{n})| + \frac{1}{2n}(|\varepsilon_{p+3N}(\tfrac{\nu}{n})| + |\varepsilon_{p+3N}(\tfrac{\nu-1}{n})|) + \frac{1}{n}|\delta_p(\tfrac{\nu}{n})|,$$

$$\begin{aligned} |\varepsilon_{q+3N}(\tfrac{\nu}{n})| \le{} & |\varepsilon_{q+3N}(\tfrac{\nu-1}{n})| + \frac{\phi}{n}(|\varepsilon_{q+1+3N}(\tfrac{\nu}{n})| + |\varepsilon_{q+1+3N}(\tfrac{\nu-1}{n})|) \\ & + \frac{\phi^2}{2n}(|\varepsilon_q(\tfrac{\nu}{n})| + |\varepsilon_q(\tfrac{\nu-1}{n})|) \\ & + \frac{\tilde{M}}{n}\sum_{s=1}^{N}(|\varepsilon_s(\tfrac{\nu}{n})| + |\varepsilon_s(\tfrac{\nu-1}{n})|) + \frac{1}{n}|\delta_{q+3N}(\tfrac{\nu}{n})|, \end{aligned}$$

$$\begin{aligned} |\varepsilon_{q+1+3N}(\tfrac{\nu}{n})| \le{} & |\varepsilon_{q+1+3N}(\tfrac{\nu-1}{n})| + \frac{\phi}{n}(|\varepsilon_{q+3N}(\tfrac{\nu}{n})| + |\varepsilon_{q+3N}(\tfrac{\nu-1}{n})|) \\ & + \frac{\phi^2}{2n}(|\varepsilon_{q+1}(\tfrac{\nu}{n})| + |\varepsilon_{q+1}(\tfrac{\nu-1}{n})|) \\ & + \frac{\tilde{M}}{n}\sum_{s=1}^{N}(|\varepsilon_s(\tfrac{\nu}{n})| + |\varepsilon_s(\tfrac{\nu-1}{n})|) + \frac{1}{n}|\delta_{q+1+3N}(\tfrac{\nu}{n})|, \end{aligned}$$

$$\begin{aligned} |\varepsilon_{q+2+3N}(\tfrac{\nu}{n})| \le{} & |\varepsilon_{q+2+3N}(\tfrac{\nu-1}{n})| + \frac{\tilde{M}}{n}\sum_{s=1}^{N}(|\varepsilon_s(\tfrac{\nu}{n})| + |\varepsilon_s(\tfrac{\nu-1}{n})|) \\ & + \frac{1}{n}|\delta_{q+2+3N}(\tfrac{\nu}{n})|. \end{aligned}$$

Adding these inequalities with respect to p=1(1)3N and q=1(1)3N-2 and applying the notations (2.3.27), we get

$$\begin{aligned} [[\varepsilon(\tfrac{\nu}{n})]]_1 \le{} & [[\varepsilon(\tfrac{\nu-1}{n})]]_1 + \frac{1}{2n}([[\varepsilon(\tfrac{\nu}{n})]]_2 + [[\varepsilon(\tfrac{\nu-1}{n})]]_2) \\ & + \frac{1}{n}[[\delta(\tfrac{\nu}{n})]]_1, \end{aligned}$$

$$\begin{aligned} [[\varepsilon(\tfrac{\nu}{n})]]_2 \le{} & [[\varepsilon(\tfrac{\nu-1}{n})]]_2 + \frac{\phi}{n}([[\varepsilon(\tfrac{\nu}{n})]]_2 + [[\varepsilon(\tfrac{\nu-1}{n})]]_2) \\ & + \frac{\phi^2}{2n}([[\varepsilon(\tfrac{\nu}{n})]]_1 + [[\varepsilon(\tfrac{\nu-1}{n})]]_1) \\ & + \frac{3N\tilde{M}}{n}([[\varepsilon(\tfrac{\nu}{n})]]_1 + [[\varepsilon(\tfrac{\nu-1}{n})]]_1) + \frac{1}{n}[[\delta(\tfrac{\nu}{n})]]_2. \end{aligned}$$

Since from the assumptions of our theorem we have $\tilde{M} \ge 1/3N$ and $\phi \ge 0$, then adding the above inequalities, we have

$$\begin{aligned} [[\varepsilon(\tfrac{\nu}{n})]] \le{} & [[\varepsilon(\tfrac{\nu-1}{n})]] + \frac{3N\tilde{M}}{n}([[\varepsilon(\tfrac{\nu}{n})]] + [[\varepsilon(\tfrac{\nu-1}{n})]]) \\ & + (\frac{\phi^2}{2n} + \frac{\phi}{n})([[\varepsilon(\tfrac{\nu}{n})]] + [[\varepsilon(\tfrac{\nu-1}{n})]]) + \frac{1}{n}[[\delta(\tfrac{\nu}{n})]] \\ ={} & [[\varepsilon(\tfrac{\nu-1}{n})]] + \frac{1}{n}(\phi + \frac{\phi^2}{2} + 3N\tilde{M})([[\varepsilon(\tfrac{\nu}{n})]] + [[\varepsilon(\tfrac{\nu-1}{n})]]) \end{aligned}$$

$$+ \frac{1}{n}[[\delta(\frac{\nu}{n})]].$$

If we denote $\alpha=\phi+\phi^2/2+3N\tilde{M}$, then this inequality can be rewritten in the form

$$(1-\frac{\alpha}{n})[[\varepsilon(\frac{\nu}{n})]] \leq (1+\frac{\alpha}{n})[[\varepsilon(\frac{\nu-1}{n})]]+\frac{1}{n}[[\delta(\frac{\nu}{n})]].$$

From our assumption we have $n>\alpha$. Therefore,

$$[[\varepsilon(\frac{\nu}{n})]] \leq \frac{1}{1-\frac{\alpha}{n}}(1+\frac{\alpha}{n})[[\varepsilon(\frac{\nu-1}{n})]]+\frac{1}{n}[[\delta(\frac{\nu}{n})]] .$$

This inequality implies

$$[[\varepsilon(\frac{\nu}{n})]] \leq \frac{(1+\frac{\alpha}{n})^{\nu}}{(1-\frac{\alpha}{n})^{\nu}}[[\varepsilon(0)]]+\frac{1}{n}\sum_{\mu=1}^{\nu}\frac{(1+\frac{\alpha}{n})^{\nu-\mu}}{(1-\frac{\alpha}{n})^{\nu+1-\mu}}[[\delta(\frac{\mu}{n})]].$$

Since $\beta_1\geq\beta_2$, we get $(1-\frac{\alpha}{n})^{\beta_1}\leq(1-\frac{\alpha}{n})^{\beta_2}$ and then $([[\varepsilon(0)]]= =[[\delta(0)]])$

$$[[\varepsilon(\frac{\nu}{n})]] \leq \frac{(1+\frac{\alpha}{n})^{n}}{(1-\frac{\alpha}{n})^{n}}([[\delta(0)]]+\frac{1}{n}\sum_{\nu=1}^{n}[[\delta(\frac{\nu}{n})]])$$

$$\leq (1+\frac{\alpha}{n})^{n}(1-\frac{\alpha}{n})^{-n}||\delta||_{E_n^o}$$

$$\leq e^{\alpha}(\frac{[\alpha]-\alpha+1}{[\alpha]+1})^{-[\alpha]-1}||\delta||_{E_n^o} .$$

The above inequality is true for each $\nu=0(1)n$. Thus our proof is finished.

Theorems 4.2 and 4.3 ensure the fulfilment of the conditions (ii) and (iii) in Th.1.2 on convergence of the discretization method. Leaving it to the reader to check the condition (i) of this theorem for our method, finally we can formulate:

Conclusion 4.1. If there exist the constants (2.3.23), then the discrete mechanics method in a rotating frame (4.3.5) is convergent on the initial value problem (4.3.6).

4.4. A MOTION NEARBY THE EQUILIBRIUM POINTS

In this and the next section we will engage in applications of discrete mechanics formulas in a rotating frame of reference. First, let us consider a problem of motion of a body with an infinitely small mass nearby the equilibrium points.

The application of the discrete formulas (4.3.2) to the differential equations of motion (4.2.2) yields the following relations

$$
\begin{aligned}
\frac{x_l^{k+1}-x_l^k}{\Delta t} &= \frac{v_l^{k+1}+v_l^k}{2}\ ; \quad l=1,2,3; \\
\frac{v_1^{k+1}-v_1^k}{\Delta t} &= v_2^{k+1}+v_2^k+\frac{1}{2}(x_1^{k+1}+x_1^k)+a_{1j} \\
&\quad -G\left[m_1\frac{x_1^{k+1}+x_1^k+2(a_{1j}+\mu)}{r_1^{k+1}r_1^k(r_1^{k+1}+r_1^k)}+m_2\frac{x_1^{k+1}+x_1^k+2(a_{1j}+\mu-1)}{r_2^{k+1}r_2^k(r_2^{k+1}+r_2^k)}\right], \\
\frac{v_2^{k+1}-v_2^k}{\Delta t} &= -v_1^{k+1}-v_1^k+\frac{1}{2}(x_2^{k+1}+x_2^k)+a_{2j} \\
&\quad -G(x_2^{k+1}+x_2^k+2a_{2j})\left[\frac{m_1}{r_1^{k+1}r_1^k(r_1^{k+1}+r_1^k)}\right. \\
&\quad \left.+\frac{m_2}{r_2^{k+1}r_2^k(r_2^{k+1}+r_2^k)}\right], \\
\frac{v_3^{k+1}-v_3^k}{\Delta t} &= -G(x_3^{k+1}+x_3^k)\left[\frac{m_1}{r_1^{k+1}r_1^k(r_1^{k+1}+r_1^k)}+\frac{m_2}{r_2^{k+1}r_2^k(r_2^{k+1}+r_2^k)}\right],
\end{aligned}
\tag{4.4.1}
$$

where $G=1/(m_1+m_2)$, $\mu=m_2/(m_1+m_2)$, a_{lj} $(l=1,2,3;\ j=1(1)5)$ denote coordinates of j-th equilibrium point and

$$
\begin{aligned}
r_1^s &= [(x_1^s+a_{1j}+\mu)^2+(x_2^s+a_{2j})^2+(x_3^s)^2]^{1/2}, \\
r_2^s &= [(x_1^s+a_{1j}+\mu-1)^2+(x_2^s+a_{2j})^2+(x_3^s)^2]^{1/2} \quad (s=k \text{ or } s=k+1).
\end{aligned}
$$

For futher simplicity let us introduce the following notations

$$
A^k = \frac{m_1}{r_1^{k+1}r_1^k(r_1^{k+1}+r_1^k)}\ , \quad B^k = \frac{m_2}{r_2^{k+1}r_2^k(r_2^{k+1}+r_2^k)}\ ,
$$

$$C_l^k = \begin{cases} 2[A^k-(1-\mu)B^k], \text{ for } l=1, \\ 0, \text{ for } l=2,3, \end{cases} \tag{4.4.2}$$
$$\tilde{F}_l^k = -G[(x_l^{k+1}+x_l^k+2a_{lj})(A^k+B^k)+C_l^k]; \; l=1,2,3,$$

where $a_{3j}=0$. Using (4.4.2), we can write the equations (4.4.1) in the form

$$v_l^{k+1} = \frac{2}{\Delta t}(x_l^{k+1}-x_l^k)-v_l^k; \quad l=1,2,3; \tag{4.4.3}$$

$$[\frac{(\Delta t)^2}{2}-2]x_1^{k+1}+2\Delta t v_1^k+[\frac{(\Delta t)^2}{2}+2]x_1^k+2\Delta t(x_2^{k+1}-x_2^k)$$
$$+(\Delta t)^2(a_{1j}+\tilde{F}_1^k) = 0,$$
$$[\frac{(\Delta t)^2}{2}-2]x_2^{k+1}+2\Delta t v_2^k+[\frac{(\Delta t)^2}{2}+2]x_2^k-2\Delta t(x_1^{k+1}-x_1^k)$$
$$+(\Delta t)^2(a_{2j}+\tilde{F}_2^k) = 0, \tag{4.4.4}$$
$$-2x_3^{k+1}+2\Delta t v_3^k+2x_3^k+(\Delta t)^2\tilde{F}_3^k = 0.$$

The equations (4.4.4) compose a system of three nonlinear equations with the unknowns x_l^{k+1} ($l=1,2,3$). In order to solve this system we can use the iteration process (1.3.43) assuming the solution obtained at the moment t_k as an initial approximation of the solution at the moment t_{k+1}.

It is easy to show (see the proof of Th.4.1) that the equations (4.4.1) (or (4.4.3)-(4.4.4)) conserve the Jacobi constant of motion, i.e. for each $k=0,1,2,...$

$$\frac{1}{2}\sum_{l=1}^{3}(v_l^k)^2 - G(\frac{m_1}{r_1^k}+\frac{m_2}{r_2^k}) - \frac{1}{2}[(x_1^k+a_{1j})^2+(x_2^k+a_{2j})^2] = C,$$

where C=const.

Below we present an algorithm with an automatic step size correction for solving the discrete equations (4.4.1). As input data in this algorithm we have:

j - equilibrium point number,

m_1,m_2 - masses of bodies,

t_o - initial moment,

Δt - basic step size of the solution,

n - number of steps ($n=(T-t_o)/\Delta t$, where T denotes the final moment),

x_l^o, v_l^o - coordinates and components of velocity at the initial moment ($l=1,2,3$),

ε - accuracy in an iteration process of calculating a_{1j} (for j=1,2 or 3),

$\bar{\varepsilon}$ - admissible discretization error,

ε_1 - accuracy in an iteration process for solving the equations (4.4.4).

As the results we get:

x_1^k, v_1^k - coordinates and components of velocity at the moment t_k (k=1(1)n).

<u>Algorithm 22</u>

1° $G := \frac{1}{m_1+m_2}$;

$\mu := m_2 G$;

$a_{3j} := 0$;

if j=4 or j=5 then:

$a_{1j} := \frac{1}{2}\frac{m_1-m_2}{m_1+m_2}$;

$a_{2j} := \sqrt{3}/2$;

if j=5 then:

$a_{2j} := -a_{2j}$;

go to 3°;

$a_{2j} := 0$;

$\alpha_3 := 6\mu^2-6\mu+1$;

$\alpha_4 := 4\mu-2$;

if j=1 then:

$\alpha_o := 2\mu^3-3\mu^2+3\mu-1$;

$\alpha_1 := \mu^4-2\mu^3+5\mu^2-4\mu+2$;

$\alpha_2 := 4\mu^3-6\mu^2+4\mu-1$;

if j=2 then:

$\alpha_o := -3\mu^2+3\mu-1$;

$\alpha_1 := \mu^4-2\mu^3+\mu^2-4\mu+2$;

$\alpha_2 := 4\mu^3-6\mu^2+2\mu-1$;

if j=3 then:

$\alpha_o := 3\mu^2-3\mu+1$;

$\alpha_1 := \mu^4-2\mu^3+\mu^2+4\mu-2$;

$\alpha_2 := 4\mu^3-6\mu^2+2\mu+1$;

$\bar{a}_{1j} := -1-\mu$;

if j=1 or j=2 then:

$\bar{a}_{1j} := 1-\mu$;

$$2^{\circ}\quad a_{1j}:=\bar{a}_{1j}-\frac{\alpha_o+\alpha_1\bar{a}_{1j}+\alpha_2\bar{a}_{1j}^2+\alpha_3\bar{a}_{1j}^3+\alpha_4\bar{a}_{1j}^4+\bar{a}_{1j}^5}{\alpha_1+2\alpha_2\bar{a}_{1j}+3\alpha_3\bar{a}_{1j}^2+4\alpha_4\bar{a}_{1j}^3+5\bar{a}_{1j}^4};$$

if $|a_{1j}-\bar{a}_{1j}|<\varepsilon$ then:

go to 3°;

$\bar{a}_{1j}:=a_{1j}$;

go to 2°;

3° $H:=\Delta t$;

$S:=$ <u>true</u>;

$k:=0$;

4° $t:=t_o+k\Delta t$;

5° for $l=1,2,3$:

$\bar{x}_l^{k+1}:=x_l^{k+1}:=x_l^k$;

$\bar{v}_l^{k+1}:=v_l^k$;

$$6^{\circ}\quad x_1^{k+1}:=x_1^{k+1}-\frac{(H^2/2-2)x_1^{k+1}+2Hv_1^k+(H^2/2+2)x_1^k+2H(x_2^{k+1}-x_2^k)+H^2(a_{1j}+\tilde{F}_1^k}{H^2/2-2+H^2\dfrac{\partial\tilde{F}_1^k}{\partial x_1^{k+1}}}$$

$$x_2^{k+1}:=x_2^{k+1}-\frac{(H^2/2-2)x_2^{k+1}+2Hv_2^k+(H^2/2+2)x_2^k-2H(x_1^{k+1}-x_1^k)+H^2(a_{2j}+\tilde{F}_2^k}{H^2/2-2+H^2\dfrac{\partial\tilde{F}_2^k}{\partial x_2^{k+1}}}$$

$$x_3^{k+1}:=x_3^{k+1}-\frac{-2x_3^{k+1}+2Hv_3^k+2x_3^k+H^2\tilde{F}_3^k}{-2+H^2\dfrac{\partial\tilde{F}_3^k}{\partial x_3^{k+1}}},$$

where $\tilde{F}_l^k=-G[(x_l^{k+1}+x_l^k+2a_{lj})(A^k+B^k)+C_l^k]$ $(l=1,2,3)$ and A^k, B^k, C_l^k are given by (4.4.2), and where

$$\frac{\partial\tilde{F}_l^k}{\partial x_l^{k+1}}=-G\left[A^k+B^k+(x_l^{k+1}+x_l^k+2a_{lj})\left(\frac{\partial A^k}{\partial x_l^{k+1}}+\frac{\partial B^k}{\partial x_l^{k+1}}\right)+\frac{\partial C_l^k}{\partial x_l^{k+1}}\right]$$

with $$\frac{\partial A^k}{\partial x_l^{k+1}}=-\frac{A^k}{r_1^{k+1}}\left[x_l^{k+1}+a_{lj}+\left(\begin{array}{l}\mu,\text{for } l=1\\ 0,\text{for } l\neq 1\end{array}\right)\right]\left(\frac{1}{r_1^{k+1}}+\frac{1}{r_1^{k+1}+r_1^k}\right)$$

$$\frac{\partial B^k}{\partial x_l^{k+1}}=-\frac{B^k}{r_2^{k+1}}\left[x_l^{k+1}+a_{lj}+\left(\begin{array}{l}\mu-1,\text{for } l=1\\ 0\ \ ,\text{for } l\neq 1\end{array}\right)\right]\left(\frac{1}{r_2^{k+1}}+\frac{1}{r_2^{k+1}+r_2^k}\right),$$

(r_1^s and r_2^s for $s=k$ or $s=k+1$ are given on page 195);

$$\frac{\partial C_l^k}{\partial x_l^{k+1}} = \begin{cases} 2[\mu \dfrac{\partial A^k}{\partial x_l^{k+1}} - (1-\mu)\dfrac{\partial B^k}{\partial x_l^{k+1}}], & \text{for } l=1, \\ 0, & \text{for } l=2,3; \end{cases}$$

for $l=1,2,3$:

    $v_l^{k+1} := \frac{2}{\Delta t}(x_l^{k+1} - x_l^k) - v_l^k$;

7° if for each $l=1,2,3$

    $|x_l^{k+1} - \bar{x}_l^{k+1}| < \varepsilon_1$ and $|v_l^{k+1} - \bar{v}_l^{k+1}| < \varepsilon_1$ then:

        go to 8°;

for $l=1,2,3$:

    $\bar{x}_l^{k+1} := x_l^{k+1}$;

    $\bar{v}_l^{k+1} := v_l^{k+1}$;

go to 6°;

8° if S=<u>true</u> then:

    go to 9°;

$H := H_1$;

S:=<u>true</u>;

printout x_l^{k+1} and v_l^{k+1} $(l=1,2,3)$;

$k := k+1$;

if $k \leq n-1$ then:

    go to 4°;

stop;

9° F:=<u>true</u>;

for $l=1,2,3$:

    $y_l := x_l^k$;

    $w_l := v_l^k$;

10° for $l=1,2,3$:

    $\bar{z}_l := z_l := y_l$;

    $\bar{u}_l := w_l$;

11° $$z_1 := z_1 - \frac{(H^2/8-2)z_1 + Hw_1 + (H^2/8+2)y_1 + H(z_2-y_2) + H^2(a_{1j}+\tilde{F}_1)/4}{H^2/8-2+\frac{H^2}{4}\frac{\partial \tilde{F}_2}{\partial z_1}};$$

$$z_2 := z_2 - \frac{(H^2/8-2)z_2 + Hw_2 + (H^2/8+2)y_2 - H(z_1-y_1) + H^2(a_{2j}+\tilde{F}_2)/4}{H^2/8-2+\frac{H^2}{4}\frac{\partial \tilde{F}_2}{\partial z_2}};$$

$$z_3 := z_3 - \frac{-2z_3 + Hw_3 + 2y_3 + H^2\tilde{F}_3/4}{-2+\frac{H^2}{4}\frac{\partial \tilde{F}_3}{\partial z_3}},$$

where $\tilde{F}_l = -G[(z_l+y_l+2a_{lj})(A+B)+C_l]$ $(l=1,2,3)$ with

$$A=\frac{m_1}{r_1\rho_1(r_1+\rho_1)}, \quad B=\frac{m_2}{r_2\rho_2(r_2+\rho_2)},$$

$$r_1=[(z_1+a_{1j}+\mu)^2+(z_2+a_{2j})^2+z_3^2]^{1/2},$$

$$\rho_1=[(y_1+a_{1j}+\mu)^2+(y_2+a_{2j})^2+y_3^2]^{1/2},$$

$$r_2=[(z_1+a_{1j}+\mu-1)^2+(z_2+a_{2j})^2+z_3^2]^{1/2},$$

$$\rho_2=[(y_1+a_{1j}+\mu-1)^2+(y_2+a_{2j})^2+y_3^2]^{1/2},$$

$$C_l=\begin{cases} 2[\mu A-(1-\mu)B], & \text{for } l=1,\\ 0, & \text{for } l=2,3,\end{cases}$$

and $$\frac{\partial \tilde{F}_l}{\partial z_l} = -G\left[A+B+(z_l+y_l+2a_{lj})\left(\frac{\partial A}{\partial z_l}+\frac{\partial B}{\partial z_l}\right)+\frac{\partial C_l}{\partial z_l}\right],$$

where $$\frac{\partial A}{\partial z_l} = -\frac{A}{r_1}\left[z_l+a_{lj}+\left(\begin{matrix}\mu, \text{for } l=1\\ 0, \text{for } l\neq 1\end{matrix}\right)\right]\left(\frac{1}{r_1}+\frac{1}{r_1+\rho_1}\right),$$

$$\frac{\partial B}{\partial z_2} = -\frac{B}{r_2}\left[z_l+a_{lj}+\left(\begin{matrix}\mu-1, & \text{for } l=1\\ 0\ , & \text{for } l\neq 1\end{matrix}\right)\right]\left(\frac{1}{r_2}+\frac{1}{r_2+\rho_2}\right),$$

$$\frac{\partial C_l}{\partial z_l} = \begin{cases} 2\left[\mu\dfrac{\partial A}{\partial z_l}-(1-\mu)\dfrac{\partial B}{\partial z_l}\right], & \text{for } l=1,\\ 0, & \text{for } l=2,3;\end{cases}$$

for $l=1,2,3$:
  $u_l := 4(z_l-y_l)/H+w_l$;

12° if for each $l=1,2,3$
  $|z_l-\bar{z}_l|<\varepsilon_1$ and $|u_l-\bar{u}_l|<\varepsilon_1$ then:
    if F=<u>false</u> then:
      go to 13°;
    F:=<u>false</u>;
    for $l=1,2,3$:
      $y_l:=z_l$;
      $w_l:=u_l$;
    go to 10°;
  for $l=1,2,3$:
    $\bar{z}_l:=z_l$;
    $\bar{u}_l:=u_l$;
  go to 11°;

13° $\tilde{h}:=H/[2 \max_{l=1,2,3}(|x_l^{k+1}-z_l|,|v_l^{k+1}-u_l|)/\bar{\varepsilon}]^{1/2}$;
  continue as in Algorithm 9, 11°.

Applying the above algorithm to determine the motion of a body with an infinitely small mass nearby the equilibrium points of the system earth-moon, we get the results given in Table XL. It has been assumed that $m_1=1$, $m_2=0.0123001$ and the mean distance between the earth and the moon, i.e. 384000 km, has been taken as a distance unit. The numbers of orbits in Table XL correspond to the numbers of orbits in Figs. 30-34.

Table XL. Solution of the equations of motion nearby the equilibrium points of the earth-moon system by the discrete mechanics method ($\Delta t=0.01$, $\varepsilon=0.0000000001$)

point	orbit number	k	l	$x_l \times 10^4$	$v_l \times 10^4$	Jacobi's const characteristics
L_1	1	0	1	1.00000000	0.75000000	C=-1.5941706067
			2	1.00000000	-7.25000000	
		60	1	0.98605953	-0.23057862	$\alpha=16$, $\beta=3$
			2	-3.04725418	-4.28256183	C as above
		120	1	2.51717941	8.13755042	$\alpha=16$, $\beta=3$
			2	-3.43336354	1.52093838	C as above
						CPU time 0.45 sec
	2	0	1	1.00000000	0.50000000	C=-1.5941706082
			2	1.00000000	-7.25000000	
		70	1	0.67351673	-0.74604555	$\alpha=16$, $\beta=3$
			2	-3.26523441	-2.44308298	C as above
		140	1	2.55893410	9.65510126	$\alpha=16$, $\beta=3$
			2	-2.16012214	2.65524380	C as above
						CPU time 0.53 sec
	3	0	1	1.00000000	0.25000000	C=-1.5941706092
			2	1.00000000	-7.25000000	
		85	1	0.13101435	-1.44860642	$\alpha=16$, $\beta=3$
			2	-3.23139725	0.50732561	C as above
		170	1	1.83159481	8.62854411	$\alpha=16$, $\beta=3$
			2	0.38814459	2.88444838	C as above
						CPU time 0.65 sec
	4	0	1	1.00000000	0.20000000	C=-1.5941706093
			2	1.00000000	-7.25000000	
		95	1	-0.12536330	-1.48552850	$\alpha=16$, $\beta=3$
			2	-3.03864494	2.26695256	C as above
		190	1	2.14660594	9.42073466	$\alpha=16$, $\beta=3$
			2	1.43490675	0.47793596	C as above
						CPU time 0.75 sec
	5	0	1	1.00000000	0.15000000	C=-1.5941706094
			2	1.00000000	-7.25000000	
		115	1	-0.56742031	-1.23337959	$\alpha=16$, $\beta=3$
			2	-2.19067308	5.25996984	C as above
		230	1	2.22333887	7.53441491	$\alpha=16$, $\beta=3$
			2	2.07168391	-5.16639078	C as above
						CPU time 0.87 sec
	6	0	1	1.00000000	0.12500000	C=-1.5941706094
			2	1.00000000	-7.25000000	
		130	1	-0.87648450	-1.03402511	$\alpha=16$, $\beta=3$
			2	-1.20607997	6.85659087	C as above
		260	1	-2.52898566	-8.65754623	$\alpha=16$, $\beta=3$
			2	3.06586972	-1.95420882	C as above
						CPU time 0.97 sec

Table XL. (cont.)

point	orbit number	k	l	$x_l \times 10^4$	$v_l \times 10^4$	Jacobi's const characteristics
L_1	7	0	1	1.00000000	0.10000000	C=-1.5941706094
			2	1.00000000	-7.25000000	
		95	1	-0.37852151	-2.21850342	α=16 , β=3
			2	-2.91444007	2.61958659	C as above
		195	1	-2.05181313	-3.04064522	α=16 , β=3
			2	3.59507276	5.50104452	C as above
						CPU time 0.77 sec
L_2	1	0	1	1.00000000	1.60000000	C=-1.5860802983
			2	1.00000000	-6.05000000	
		80	1	1.08969616	-0.63583695	α=16 , β=3
			2	-3.67199252	-3.58631276	C as above
		160	1	2.41852249	6.44268461	α=16 , β=3
			2	-3.97949789	1.04079884	C as above
						CPU time 0.66 sec
	2	0	1	1.00000000	1.20000000	C=-1.5860803039
			2	1.00000000	-6.05000000	
		105	1	0.22286547	-1.54185345	α=13 , β=3
			2	-3.75313765	-0.01991336	C as above
		210	1	1.94294987	7.54818395	α=16 , β=3
			2	-0.40203563	1.82673668	C as above
						CPU time 0.81 sec
	3	0	1	1.00000000	1.10000000	C=-1.5860803051
			2	1.00000000	-6.05000000	
		125	1	-0.33187065	-1.60027191	α=15 , β=3
			2	-3.33771565	2.54029049	C as above
		250	1	2.47085754	8.40513739	α=16 , β=3
			2	1.45285539	-1.54554645	C as above
						CPU time 0.93 sec
	4	0	1	1.00000000	1.05000000	C=-1.5860803056
			2	1.00000000	-6.05000000	
		145	1	-0.81115197	-1.36464283	α=15 , β=3
			2	-2.48408155	4.68707170	C as above
		290	1	2.25776954	6.03059015	α=16 , β=3
			2	2.14927170	-4.86453313	C as above
						CPU time 1.09 sec
	5	0	1	1.00000000	1.00000000	C=-1.5860803061
			2	1.00000000	-6.05000000	
		115	1	-0.39899114	-2.31868308	α=15 , β=3
			2	-3.37569588	1.79646007	C as above
		230	1	-1.72346390	-0.28321844	α=16 , β=3
			2	3.17247837	5.66936737	C as above
						CPU time 0.90 sec
	6	0	1	1.00000000	0.85000000	C=-1.5860803075
			2	1.00000000	-6.05000000	

Table XL. (cont.)

point	orbit number	k	l	$x_1 \times 10^4$	$v_1 \times 10^4$	Jacobi's const characteristics
L_2	6	80	1	0.23010920	-2.47582280	α=15 , β=3
			2	-3.15128203	-2.15704068	C as above
		160	1	-2.20991579	-3.42854929	α=15 , β=3
			2	-0.87015750	7.40215082	C as above
						CPU time 0.66 sec
L_3	1	0	1	1.00000000	0.75000000	C=-1.5060735977
			2	1.00000000	-2.00000000	
		130	1	1.27632441	0.46588018	α=30 , β=5
			2	-0.42887312	-2.50101886	C as above
		260	1	1.68594876	0.18801716	α=30 , β=5
			2	-3.86284373	-3.15251877	C as above
						CPU time 1.50 sec
	2	0	1	1.00000000	0.50000000	C=-1.5060735992
			2	1.00000000	-2.00000000	
		145	1	0.98828624	-0.57334131	α=18 , β=3
			2	-1.70661133	-1.95700871	C as above
		290	1	0.21938063	-1.04481129	α=27 , β=3
			2	-4.30987966	-0.30555132	C as above
						CPU time 1.52 sec
	3	0	1	1.00000000	0.50000000	C=-1.5060735880
			2	1.00000000	-2.50000000	
		60	1	-2.88653497	-0.79949440	α=30 , β=5
			2	2.88242563	5.28566261	C as above
						CPU time 0.40 sec
	4	0	1	1.00000000	0.50000000	C=-1.5060735939
			2	1.00000000	-2.25000000	
		70	1	-2.03729590	0.04490817	α=30 , β=5
			2	2.62606172	3.84307222	C as above
						CPU time 0.44 sec
L_4	1	0	1	1.00000000	2.20000000	C=-1.4939984892
			2	1.00000000	-1.45000000	
		150	1	2.48373642	0.61293010	α=20 , β=3
			2	-0.64028703	-1.59678222	C as above
		300	1	1.80874323	-1.29289318	α=20 , β=3
			2	-1.51403572	-0.63336379	C as above
						CPU time 1.43 sec
	2	0	1	1.00000000	2.00000000	C=-1.4939984934
			2	1.00000000	-1.45000000	
		150	1	3.36530959	-0.23088723	α=20 , β=3
			2	-1.81769790	-0.96632404	C as above
		300	1	-0.83752435	-1.29012678	α=17 , β=3
			2	-0.21973356	-0.01785606	C as above
						CPU time 1.32 sec

Table XL. (cont.)

point	orbit number	k	l	$x_l \times 10^4$	$v_l \times 10^4$	Jacobi's const characteristics
L_5	1	0	1	1.00000000	0.75000000	C=-1.4939984955
			2	1.00000000	-0.20000000	
		150	1	2.28267277	-0.96317462	α=20 , β=3
			2	0.43899857	-1.00736619	C as above
		300	1	-1.81613146	-1.11453625	α=20 , β=3
			2	-1.72051600	-0.33451353	C as above
						CPU time 1.33 sec
	2	0	1	1.00000000	0.95000000	C=-1.4939984938
			2	1.00000000	-0.20000000	
		150	1	1.40088713	-0.11904070	α=20 , β=3
			2	0.48620644	-0.88278101	C as above
		300	1	0.83016645	-1.11742037	α=20 , β=3
			2	-0.37039010	-0.91461620	C as above
						CPU time 1.33 sec
	3	0	1	1.00000000	1.15000000	C=-1.4939984917
			2	1.00000000	-0.20000000	
		150	1	0.51791031	0.72465581	α=20 , β=3
			2	0.53240373	-0.71611990	C as above
		300	1	3.47585041	-1.12020163	α=20 , β=3
			2	0.98028189	-1.49486254	C as above
						CPU time 1.33 sec

Remark! Δt=0.05 for L_3, L_4 and L_5.

4.5. DISCRETE HILL'S EQUATIONS

In order to work out discrete Hill's equations basing ourselves on discrete mechanics, let us rewrite the equations (4.1.30) as follows

$$\begin{aligned} \ddot{x}_1-2M\dot{x}_2-M^2x_1 &= F_1, \\ \ddot{x}_2+2M\dot{x}_1-M^2x_2 &= F_2, \\ \ddot{x}_3 &= F_3, \end{aligned} \tag{4.5.1}$$

where the dot denotes differentiation with respect to the time τ determined by (4.1.29) and

$$\begin{aligned} F_1 &= (2M^2-\frac{\varkappa}{r^3})x_1, \\ F_2 &= (-M^2-\frac{\varkappa}{r^3})x_2, \end{aligned} \tag{4.5.2}$$

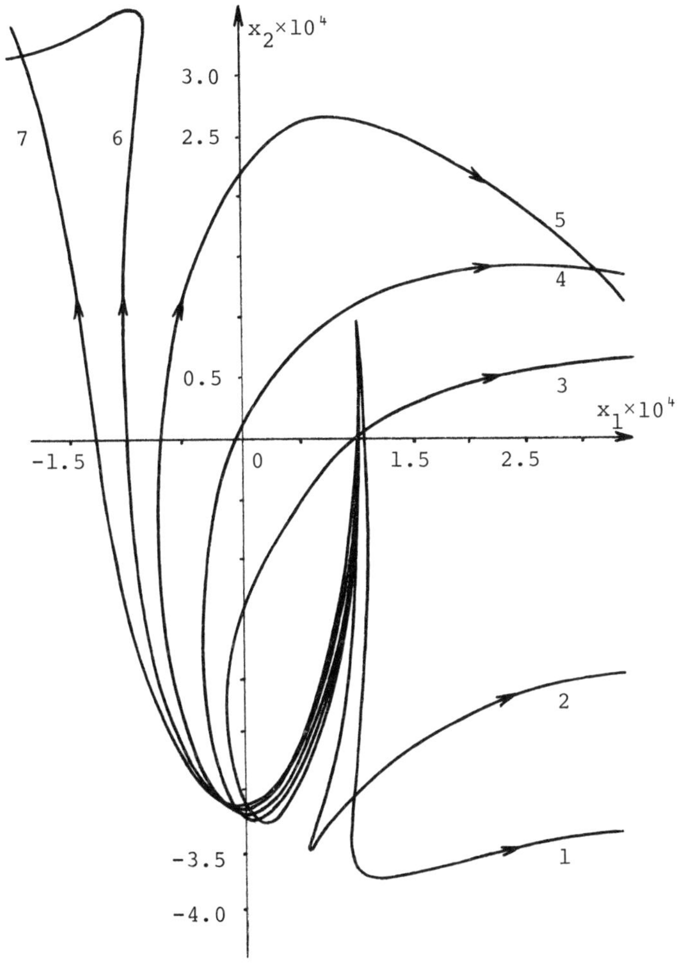

Fig.30. The orbits nearby the point L_1

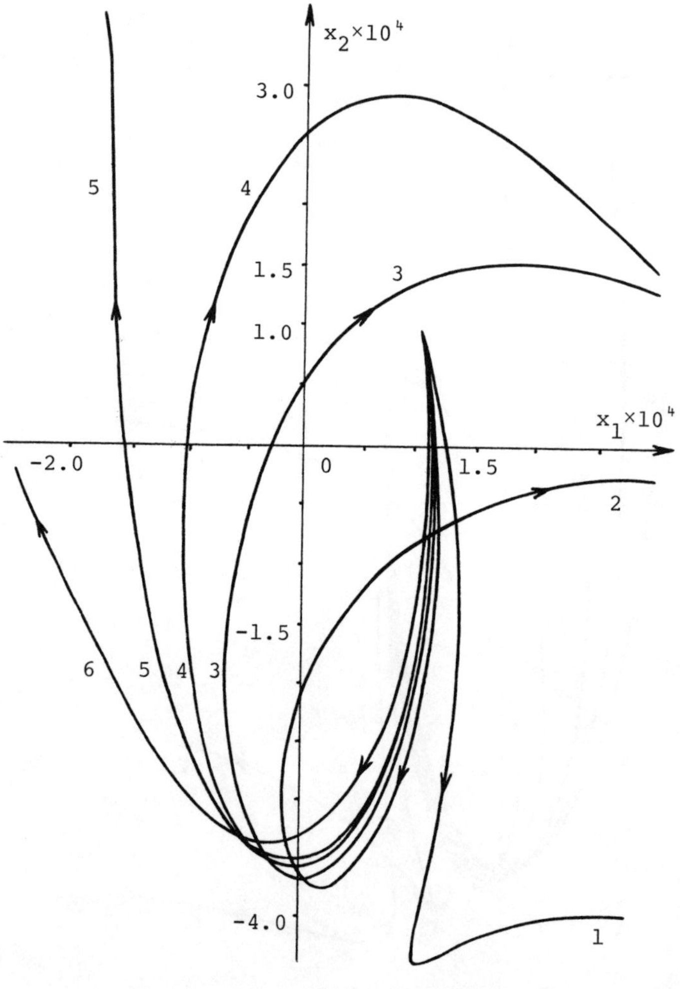

Fig.31. The orbits nearby the point L_2

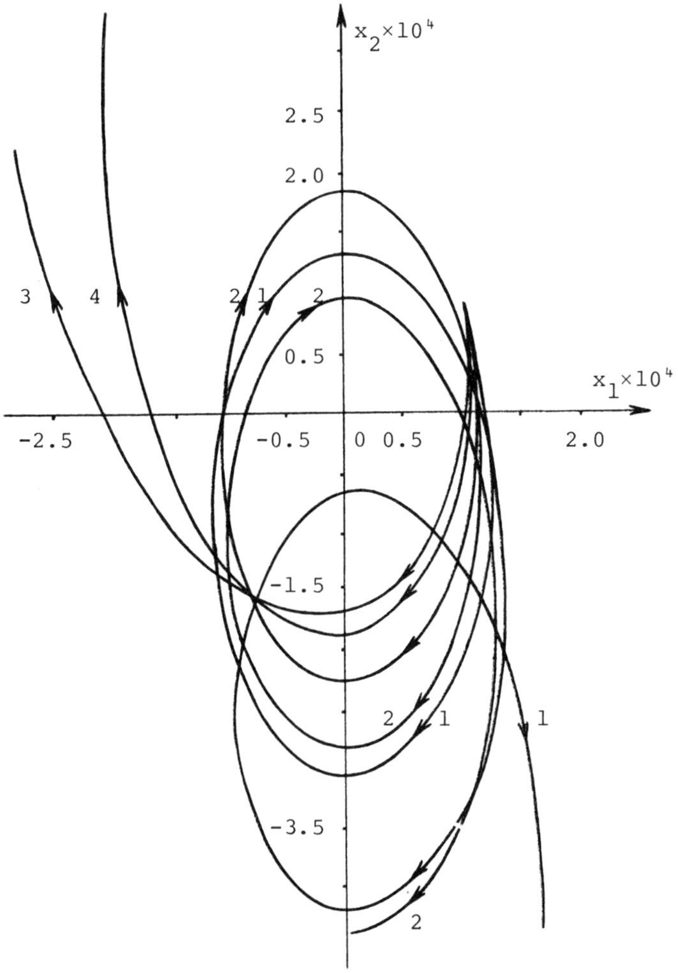

Fig.32. The orbits nearby the point L_3

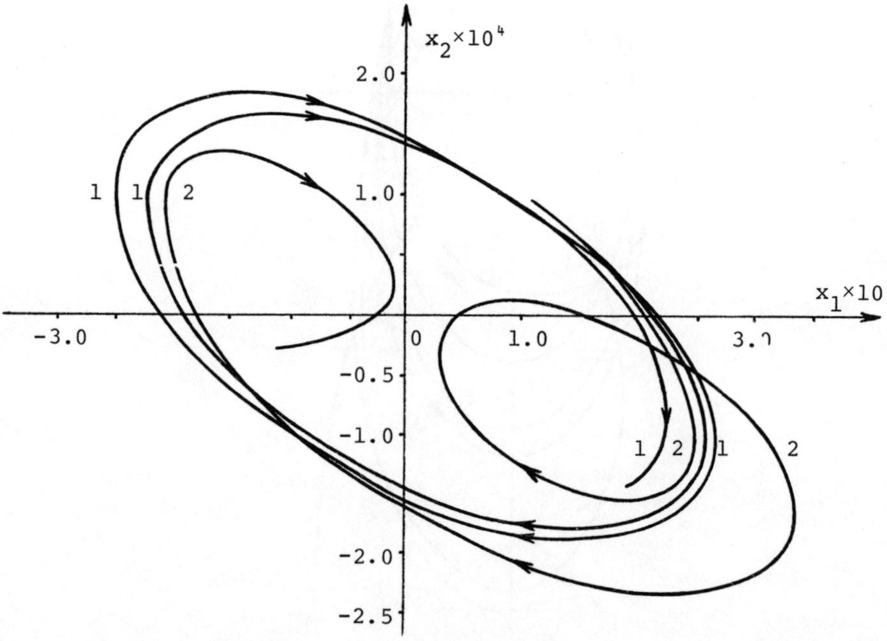

Fig.33. The orbits nearby the point L_4

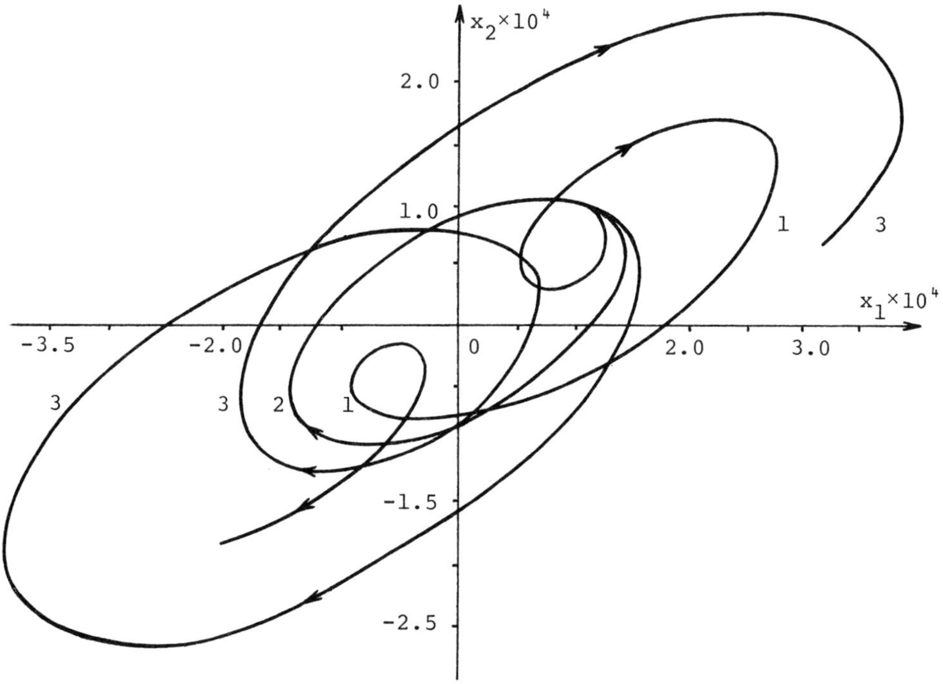

Fig.34. The orbits nearby the point L_5

$$\bar{F}_3 = (-M^2 - \frac{\kappa}{r^3})x_3,$$
$$r = (x_1^2+x_2^2+x_3^2)^{1/2}, \quad M,\kappa=\text{const.}$$

The form of equations (4.5.1) is analogous to (4.1.5). Thus we can discretiate these equations by the same method as in Sect.4.3. We get the following discrete equations

$$\begin{aligned}
\frac{x_l^{k+1}-x_l^k}{\Delta\tau} &= \frac{v_l^{k+1}+v_l^k}{2}; \quad l=1,2,3;\\
\frac{v_1^{k+1}-v_1^k}{\Delta\tau} &= M(v_2^{k+1}+v_2^k) + \frac{M^2}{2}(x_1^{k+1}+x_1^k)+\bar{F}_1^k,\\
\frac{v_2^{k+1}-v_2^k}{\Delta\tau} &= -M(v_1^{k+1}+v_1^k) + \frac{M^2}{2}(x_2^{k+1}+x_2^k)+\bar{F}_2^k,\\
\frac{v_3^{k+1}-v_3^k}{\Delta\tau} &= \bar{F}_3^k,
\end{aligned} \tag{4.5.3}$$

where

$$\begin{aligned}
\Delta\tau &= \tau_{k+1}-\tau_k,\\
\bar{F}_1^k &= [M^2 - \frac{\kappa}{r^{k+1}r^k(r^{k+1}+r^k)}](x_1^{k+1}+x_1^k),\\
\bar{F}_p^k &= [-\frac{M^2}{2} - \frac{\kappa}{r^{k+1}r^k(r^{k+1}+r^k)}](x_p^{k+1}+x_p^k); \quad p=2,3.
\end{aligned} \tag{4.5.4}$$

The formulas (4.5.4) result from (4.5.2) immediately because it is sufficient to take into account that

$$[r(t_k)]^3 \equiv (r^k)^3 = r^{k+1}r^k(r^{k+1}+r^k)+O(\Delta\tau),$$

$$x_p(t_k) \equiv x_p^k = \frac{x_p^{k+1}+x_p^k}{2} + O(\Delta\tau); \quad p=1,2,3.$$

The equations (4.5.3)-(4.5.4) are discrete Hill's equations of motion. They conserve the Jacobi constant (4.1.31). Namely, we have:

<u>Theorem 4.4.</u> For discrete Hill's equations (4.5.3)-(4.5.4) the expresion

$$\sum_{l=1}^{3} (v_l^k)^2 - 2\frac{\kappa}{r^k} - M^2[3(x_1^k)^2-(x_3^k)^2]$$

is constant for each k=0,1,2,... .

Proof. It is sufficient to show that for each k=0,1,2,... we have

$$\sum_{l=1}^{3}[(v_l^{k+1})^2-(v_l^k)^2] = M^2\{3[(x_1^{k+1})^2-(x_1^k)^2]-[(x_3^{k+1})^2-(x_3^k)^2]\}$$
$$+2\kappa\left(\frac{1}{r^{k+1}}-\frac{1}{r^k}\right).$$

From (4.5.3) and (4.5.4) we get

$$\sum_{l=1}^{3}[(v_l^{k+1})^2-(v_l^k)^2] = \sum_{l=1}^{3}(v_l^{k+1}-v_l^k)(v_l^{k+1}+v_l^k)$$
$$= \sum_{l=1}^{3} 2\,\frac{v_l^{k+1}-v_l^k}{\Delta\tau}(x_l^{k+1}-x_l^k)$$
$$= 2\{[M(v_1^{k+1}+v_1^k)+\frac{3}{2}M^2(x_1^{k+1}+x_1^k)-\frac{\kappa(x_1^{k+1}+x_1^k)}{r^{k+1}r^k(r^{k+1}+r^k)}]$$
$$\times(x_1^{k+1}-x_1^k)$$
$$+[-M(v_1^{k+1}+v_1^k)-\frac{\kappa(x_2^{k+1}+x_2^k)}{r^{k+1}r^k(r^{k+1}+r^k)}](x_2^{k+1}-x_2^k)$$
$$+[-\frac{M^2}{2}(x_3^{k+1}+x_3^k)-\frac{\kappa(x_3^{k+1}+x_3^k)}{r^{k+1}r^k(r^{k+1}+r^k)}](x_3^{k+1}-x_3^k)\}$$
$$= 2\{[\frac{2M}{\Delta\tau}(x_2^{k+1}-x_2^k)+\frac{3}{2}M^2(x_1^{k+1}+x_1^k)](x_1^{k+1}-x_1^k)$$
$$-\frac{2M}{\Delta\tau}(x_1^{k+1}-x_1^k)(x_2^{k+1}-x_2^k)-\frac{M^2}{2}(x_3^{k+1}+x_3^k)(x_3^{k+1}-x_3^k)$$
$$-\frac{\kappa}{r^{k+1}r^k(r^{k+1}+r^k)}[(x_1^{k+1})^2-(x_1^k)^2+(x_2^{k+1})^2-(x_2^k)^2$$
$$+(x_3^{k+1})^2-(x_3^k)^2]\}$$
$$= 3M^2[(x_1^{k+1})^2-(x_1^k)^2]-M^2[(x_3^{k+1})^2-(x_3^k)^2]$$
$$-2\,\frac{\kappa}{r^{k+1}r^k(r^{k+1}+r^k)}[(r^{k+1})^2-(r^k)^2]$$
$$= M^2\{3[(x_1^{k+1})^2-(x_1^k)^2]-[(x_3^{k+1})^2-(x_3^k)^2]\}+2\kappa\left(\frac{1}{r^{k+1}}-\frac{1}{r^k}\right)$$

and it brings our proof to an end.

Below we present an algorithm with an automatic step size correction for solving discrete Hill's equations. In this algorithm we have the following input data:

G - gravitational constant,

m_o - mass of the earth,

m_1 - mass of the moon,

ν - mean motion of the moon,

ν' - mean motion of the sun,

Δt - basic step size of the solution (do not mistake Δt for $\Delta\tau$ else),

t_o - initial moment,

n - the numbers of integration steps,

x_l^o, v_l^o - coordinates and components of velocity at the moment t_o $(l=1,2,3)$,

$\underline{\varepsilon}$ - accuracy in an iteration process for solving (4.5.3),

$\bar{\varepsilon}$ - admissible discretization error.

As the results we obtain:

x_l^k, v_l^k - coordinates and components of velocity at the moments $t_k=t_o+k\Delta t$ $(k=1(1)n)$.

<u>Algorithm 23.</u>

1° $\kappa := G \dfrac{m_o+m_1}{(\nu-\nu')^2}$;

$M := \dfrac{\nu'}{\nu-\nu'}$;

$\Delta\tau := (\nu-\nu')\Delta t$;

2° $H := \Delta\tau$;

S:=<u>true</u>;

k:=0;

3° $\tau := k\Delta\tau$;

4° for l=1,2,3:

$\bar{x}_l^{k+1} := x_l^{k+1} := x_l^k$;

$\bar{v}_l^{k+1} := v_l^k$;

5° $$x_1^{k+1} := x_1^{k+1} - \frac{[(MH)^2/2-2]x_1^{k+1}+2Hv_1^k+[(MH)^2/2+2]x_1^k+2MH(x_2^{k+1}-x_2^k)+H^2\bar{F}}{(MH)^2/2-2+H^2 \dfrac{\partial \bar{F}_1^k}{\partial x_1^{k+1}}}$$

$$x_2^{k+1} := x_2^{k+1} - \frac{[(MH)^2/2-2]x_2^{k+1}+2Hv_2^k+[(MH)^2/2+2]x_2^k-2MH(x_1^{k+1}-x_1^k)+H^2\bar{F}}{(MH)^2/2-2+H^2 \dfrac{\partial \bar{F}_2^k}{\partial x_2^{k+1}}}$$

$$x_3^{k+1} := x_3^{k+1} - \frac{-2x_3^{k+1}+2Hv_3^k+2x_3^k+H^2\bar{F}_3^k}{-2+H^2 \dfrac{\partial \bar{F}_3^k}{\partial x_3^{k+1}}};$$

where $\bar{F}_p^k$ $(p=1,2,3)$ is given by (4.5.4) and

$$\frac{\partial \bar{F}_1^k}{\partial x_1^{k+1}} = M^2+\gamma_1^k; \quad \frac{\partial \bar{F}_p^k}{\partial x_p^{k+1}} = -\frac{M^2}{2}+\gamma_p^k \quad (p=2,3);$$

$$\gamma_s^k = \frac{\kappa}{r^{k+1}r^k(r^{k+1}+r^k)}\left[-1+\frac{x_s^{k+1}(x_s^{k+1}+x_s^k)}{r^{k+1}}\left(\frac{1}{r^{k+1}}+\frac{1}{r^{k+1}+r^k}\right)\right]$$

(s=1,2,3);

for l=1,2,3:

$$v_l^{k+1} := \frac{2}{\Delta\tau}(x_l^{k+1}-x_l^k)-v_l^k;$$

6°-9°, 11°, 12° - as 7°-10°, 12°, 13° in Algorithm 22;

10° $$z_1 := z_1 - \frac{[(MH)^2/8-2]z_1+Hw_1+[(MH)^2/8+2]y_1+MH(z_2-y_2)+H^2\bar{F}_1/4}{(MH)^2/8-2+\frac{H^2}{4}\frac{\partial\bar{F}_1}{\partial z_1}};$$

$$z_2 := z_2 - \frac{[(MH)^2/8-2]z_2+Hw_2+[(MH)^2/8+2]y_2-MH(z_1-y_1)+H^2\bar{F}_2/4}{(MH)^2/8-2+\frac{H^2}{4}\frac{\partial\bar{F}_2}{\partial z_2}};$$

$$z_3 := z_3 - \frac{-2z_3+Hw_3+2y_3+H^2\bar{F}_3/4}{-2+\frac{H^2}{4}\frac{\partial\bar{F}_3}{\partial z_3}};$$

where $\bar{F}_1 = [M^2-\frac{\kappa}{r\rho(r+\rho)}](z_1+y_1)$;

$$\bar{F}_p = [-\frac{M^2}{2}-\frac{\kappa}{r\rho(r+\rho)}](z_p+y_p) \quad (p=2,3);$$

$$r=(\sum_{l=1}^{3} z_l^2)^{1/2}, \quad \rho=(\sum_{l=1}^{3} y_l^2)^{1/2},$$

and where

$$\frac{\partial\bar{F}_1}{\partial z_1} = M^2+\gamma_1, \quad \frac{\partial\bar{F}_p}{\partial z_p} = -\frac{M^2}{2}+\gamma_p \quad (p=2,3),$$

$$\gamma_s = \frac{\kappa}{r\rho(r+\rho)}\left[-1+\frac{z_s(z_s+y_s)}{r}\left(\frac{1}{r}+\frac{1}{r+\rho}\right)\right] \quad (s=1,2,3);$$

for l=1,2,3:

$$u_l := 4(z_l-y_l)/H+w_l.$$

The application of the above algorithm for the same data as in Sect.4.2 yields the results given in Table XLI. The Jacobi constant of motion has the same value at each moment, i.e. is equal to -1.3836049279.

The obtained Hill's orbit of the moon is presented in Fig.35.

Table XLI. Solution of discrete Hill's equations ($\Delta\tau$=0.0531922, ε=0.00000001)

k	l	x_l	v_l	characteristics
40	1	0.8217671321	0.5038909962	α=111 , β=19
	2	-0.4047968516	0.9607772789	
	3	0.0718364757	0.0479234027	
80	1	-0.1580998963	-0.9249540302	α=88 , β=15
	2	1.0211063035	-0.1431878526	
	3	-0.0300297829	-0.0852744863	
120	1	-0.6952820026	0.7323824969	α=100 , β=17
	2	-0.6215701662	-0.7689476235	
	3	-0.0371020284	0.0913023411	
160	1	0.9689253282	0.0549105105	α=99 , β=17
	2	0.0115155116	1.1211702810	
	3	0.0724393253	-0.0478203076	
200	1	-0.4442600494	-0.8392868491	α=88 , β=15
	2	0.9039023564	-0.4826801840	
	3	-0.0843609257	-0.0293097620	

CPU time 3.23 sec

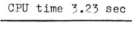

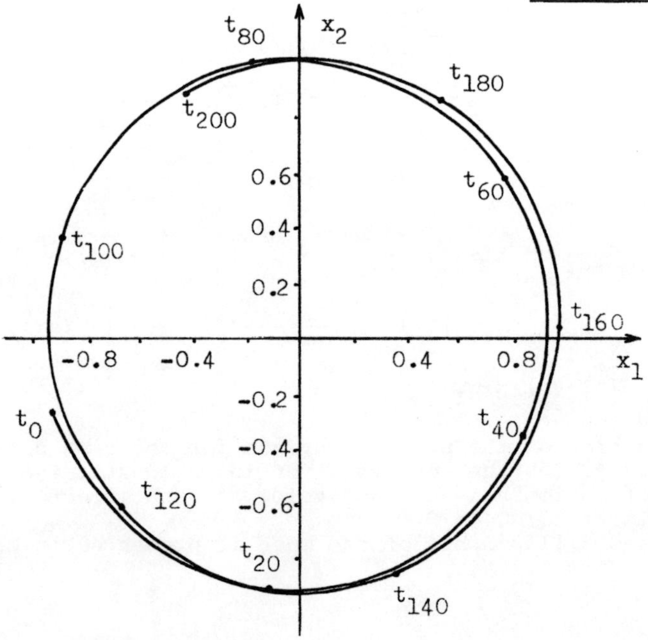

Fig.35. The orbit obtained from discrete Hill's equations

Appendix

PL/I Procedures

Procedure NBPDMG - Solution of the N-Body Problem by the Discrete Mechanics of Greenspan

1. Application

The procedure NBPDMG integrates the equations of motion (2.1.1)-(2.1.2) of N material points in the inertial frame of reference in the interval $[t_k, t_{k+1}]$ (k=0,1,2,...) by the Greenspan discrete mechanics method (2.3.35)-(2.3.36) with an automatic step size correction.

2. Data

G - gravitational constant;
TK,TK1 - ends of the integration interval $[t_k, t_{k+1}]$;
H - initial integration step size in the interval $[t_k, t_{k+1}]$ (for k=0 it can be accepted $H=t_1-t_o$);
N - number of material points;
M(1:N)- masses of material points;
X(1:3,1:N),V(1:3,1:N) - coordinates and components of velocities of N material points at the moment t_k;
EPS - admissible discretization error;
EPS1 - accuracy in the iteration process for solving the equations (2.3.36).

3. Results

H - optimal step size for beginning of integration in the next interval;
X(1:3,1:N),V(1:3,1:N) - coordinates and components of velocities of N material points at the moment t_{k+1}.

4. Procedure

```
NBPDMG: PROC (G,TK,TK1,H,N,M,X,V,EPS,EPS1);
 DCL (A,B,G,H,HV,H1,EPS,EPS1,M(*),SUM,SUM1,R,R1,T,TK,TK1,
      V(*,*),X(*,*),(U(3,N),UP(3,N),VP(3,N),V1(3,N),W(3,N),
      XP(3,N),X1(3,N),Y(3,N),Z(3,N),ZP(3,N)) CTL) FLOAT (16),
```

```
     (S,F) BIT (1),
     P BIN FIXED;
 ALLOC U,UP,VP,V1,W,XP,X1,Y,Z,ZP;
 S='1'B; T=TK;
LAB1: XP,X1=X; VP=V;
LAB2: DO L=1,2,3; DO I=1 TO N;
       A=X1(L,I)+X(L,I); SUM,SUM1=0;
       DO J=1 TO I-1,I+1 TO N;
        R,R1=0;
        DO P=1,2,3;
         R=R+(X(P,I)-X(P,J))**2; R1=R1+(X1(P,I)-X1(P,J))**2;
        END;
        R=SQRT(R); R1=SQRT(R1); B=X1(L,J)+X(L,J);
        SUM=SUM+M(J)*(A-B)/(R*R1*(R+R1));
        SUM1=SUM1+M(J)*(1-(A-B)*(X1(L,I)-X1(L,J))*(1/R1
             +1/(R+R1))/R1)/(R*R1*(R+R1));
       END;
       X1(L,I)=X1(L,I)-(X1(L,I)-X(L,I)-H*(V(L,I)-H*G*SUM/2))
               /(1+H*H*G*SUM1/2);
     END; END;
 V1=2*(X1-X)/H-V;
 DO L=1,2,3; DO I=1 TO N;
  IF ABS(X1(L,I)-XP(L,I))>=EPS1 | ABS(V1(L,I)-VP(L,I))>=EPS1
   THEN DO;
         XP=X1; VP=V1; GO TO LAB2;
        END;
 END; END;
 IF S='0'B THEN GO TO LAB5;
 F='1'B; Y=X; W=V;
LAB3: ZP,Z=Y; UP=W;
LAB4: DO L=1,2,3; DO I=1 TO N;
       A=Z(L,I)+Y(L,I); SUM,SUM1=0;
       DO J=1 TO I-1,I+1 TO N;
        R,R1=0;
        DO P=1,2,3;
         R=R+(Y(P,I)-Y(P,J))**2;
         R1=R1+(Z(P,I)-Z(P,J))**2;
        END;
        R=SQRT(R); R1=SQRT(R1); B=Z(L,J)+Y(L,J);
        SUM=SUM+M(J)*(A-B)/(R*R1*(R+R1));
        SUM1=SUM1+M(J)*(1-(A-B)*(Z(L,I)-Z(L,J))*(1/R1
             +1/(R+R1))/R1)/(R*R1*(R+R1));
       END;
       Z(L,I)=Z(L,I)-(Z(L,I)-Y(L,I)-H*(W(L,I)-H*G*SUM/4)/2)
              /(1+H*H*G*SUM1/8);
     END; END;
 U=4*(Z-Y)/H-W;
 DO L=1,2,3; DO I=1 TO N;
  IF ABS(Z(L,I)-ZP(L,I))>=EPS1 | ABS(U(L,I)-UP(L,I))>=EPS1
   THEN DO;
         ZP=Z; UP=U; GO TO LAB4;
        END;
 END; END;
```

```
 IF F THEN DO;
           F='0'B; Y=Z; W=U; GO TO LAB3;
          END;
 A=0;
 DO L=1,2,3; DO I=1 TO N;
  A=MAX(A,ABS(X1(L,I)-Z(L,I)),ABS(V1(L,I)-U(L,I)));
 END; END;
 HV=H/SQRT(2*A/EPS);
 IF HV<=H/3 THEN DO;
                 H=2*HV; GO TO LAB1;
                END;
 IF HV>=H THEN DO;
               IF T+2*H<=TK1 THEN DO;
                                   H=2*H; GO TO LAB1;
                                  END;
               H1=H; H=TK1-T; S='0'B; GO TO LAB1;
              END;
 X=Z; V=U; T=T+H;
 IF T+2 HV<TK1 THEN DO;
                    H=2*HV; GO TO LAB1;
                   END;
 H=TK1-T; H1=2*HV; S='0'B; GO TO LAB1;
LAB5: H=H1; X=X1; V=V1;
END NBPDMG;
```

Procedure NBPDMA - Solution of the N-Body Problem by the Discrete Mechanics of Arbitrary Order

1. Application

The procedure NBPDMA integrates the equations of motion (2.1.1)-(2.1.2) of N material points in the inertial frame of reference in the time interval $[t_k, t_{k+1}]$ (k=0,1,...) by the discrete mechanics of arbitrary order (see Sect.2.4).

2. Data

P - desired order of method;
G - gravitational constant;
TK,TK1 - ends of the integration interval $[t_k, t_{k+1}]$;
N - number of material points;
M(1:N) - masses of material points;
X(1:3,1:N),V(1:3,1:N) - coordinates and components of velocities of N material points at the moment t_k;
EPS1 - accuracy in the iteration process for solving the equations (2.4.1)-(2.4.2);
EPS2 - accuracy in the iteration process for solving the equation (2.4.13).

3. Results

X(1:3,1:N),V(1:3,1:N) - coordinates and components of velocities of N material points at the moment t_{k+1}.

4. Procedure

```
NBPDMA: PROC (P,G,TK,TK1,N,M,X,V,EPS1,EPS2);
 DCL (BETA(0:P) CTL,P,Q) BIN FIXED,
     (A,AP,A1,B,C,DF,EPS1,EPS2,EO,FF,G,H,M(*),R,R1,R2,SUM,
      SUM1,TK,TK1,V(*,*),X(*,*),(ALFA1(I),ALFA2(I),BK(I),
      F(L),F1(L),U(3,N),VK(I,3,N),W(3,N),WP(3,N),XK(I,3,N),
      Y(3,N),YP(3,N),Z(3,N)) CTL) FLOAT (16);
 ALLOC BETA,U,W,WP,Y,YP,Z;
 I=P+1; L=P-1;
 ALLOC ALFA1,ALFA2,BK,F,F1,VK,XK;
 BETA(0)=1; BETA(1)=2; BETA(2)=3; EO=0;
 DO K=3 TO P;
  BETA(K)=2 BETA(K-2);
 END;
 DO I=1 TO N;
  SUM=0;
  DO J=1 TO I-1,I+1 TO N;
   SUM=SUM+M(J)/SQRT((X(1,I)-X(1,J))**2+(X(2,I)-X(2,J))**2
       +(X(3,I)-X(3,J))**2);
  END;
  EO=EO+M(I)*( V(1,I)**2+V(2,I)**2+V(3,I)**2-G*SUM);
 END;
 EO=EO/2; H=TK1-TK; BK=1;
 DO K=1 TO P+1; DO L=1 TO K-1,K+1 TO P+1;
  BK(K)=BK(K)*BETA(L-1);
 END; END;
 F,F1=1;
 DO L=1 TO P-1;
  DO I=1 TO L-1,L+1 TO P-1;
   A=BK(L)-BK(I); F(L)=F(L)*(BK(P)-BK(I))/A;
   F1(L)=F1(L)*(BK(P+1)-BK(I))/A;
  END;
  F(L)=BK(P)*F(L)/BK(L); F1(L)=BK(P+1)*F1(L)/BK(L);
 END;
 A,C=0;
 DO L=1 TO P-1;
  A=A+F1(L); C=C+F(L);
 END;
 A=1-A; C=1-C; ALFA1(P+1)=0; ALFA1(P)=1/C; ALFA2(P+1)=1;
 ALFA2(P)=-A/C;
 DO L=1 TO P-1;
  ALFA1(L)=-F(L)/C; ALFA2(L)=A*F(L)/C-F1(L);
 END;
 SUM,SUM1=0;
 DO K=1 TO P+1;
  A=BETA(K-1)**P; SUM=SUM+ALFA1(K)/A; SUM1=SUM1+ALFA2(K)/A;
 END;
 A1=-SUM/SUM1;
 DO K=1 TO P+1;
  Z=X; U=V; MI=0;
LAB1: YP,Y=Z; WP=U;
```

```
LAB2: DO L=1,2,3; DO I=1 TO N;
       SUM,SUM1=0; A=Y(L,I)+Z(L,I);
       DO J=1 TO I-1,I+1 TO N;
        R,R1=0;
        DO Q=1,2,3;
         R=R+(Z(Q,I)-Z(Q,J))**2; R1=R1+(Y(Q,I)-Y(Q,J))**2;
        END;
        R=SQRT(R); R1=SQRT(R1); B=Y(L,J)+Z(L,J);
        SUM=SUM+M(J)*(A-B)/(R*R1*(R+R1));
        SUM1=SUM1+M(J)*(1-(A-B)*(Y(L,I)-Y(L,J))+(1/R1
             +1/(R+R1))/R1)/(R*R1*(R+R1));
       END;
       Y(L,I)=Y(L,I)-(Y(L,I)-Z(L,I)-H*(U(L,I)-H*G*SUM
              /(2*BETA(K-1)))/BETA(K-1))/(1+H*H*G*SUM1
              /(2*BETA(K-1)**2));
      END; END;
  W=2*BETA(K-1)*(Y-Z)/H-U;
  DO L=1,2,3; DO I=1 TO N;
   IF ABS(Y(L,I)-YP(L,I))>=EPS1 | ABS(W(L,I)-WP(L,I))>=EPS1
    THEN DO;
          YP=Y; WP=W; GO TO LAB2;
         END;
  END; END;
  MI=MI+1;
  IF MI<=BETA(K-1)-1 THEN DO;
                            Z=Y; U=W; GO TO LAB1;
                           END;
  XK(K,*,*)=Y; VK(K,*,*)=W;
 END;
 AP=A1;
LAB3: FF,DF=0;
 DO I=1 TO N;
  SUM,SUM1=0;
  DO J=1 TO I-1,I+1 TO N;
   R,C=0;
   DO L=1,2,3;
    R1,R2=0;
    DO K=1 TO P+1;
     B=XK(K,L,I)-XK(K,L,J); R1=R1+(ALFA1(K)+ALFA2(K)*A1)*B;
     R2=R2+ALFA2(K)*B;
    END;
    R=R+R1**2; C=C+R1*R2;
   END;
   SUM=SUM+M(J)/SQRT(R); SUM1=SUM1+M(J)*C/R**1.5;
  END;
  R,C=0;
  DO L=1,2,3;
   R1,R2=0;
   DO K=1 TO P+1;
    B=VK(K,L,I); R1=R1+(ALFA1(K)+ALFA2(K)*A1)*B;
    R2=R2+ALFA2(K)*B;
   END;
```

```
   R=R+R1**2; C=C+R1*R2;
  END;
  FF=FF+M(I)*(R-G*SUM); DF=DF+M(I)*(2*C+G*SUM1);
 END;
 A1=A1-(FF-2*EO)/DF;
 IF ABS(A1-AP)>=EPS2 THEN DO;
                            AP=A1; GO TO LAB3;
                           END;
 X,V=0;
 DO K=1 TO P+1;
  B=ALFA1(K)+ALFA2(K)*A1; X=X+B*XK(K,*,*); V=V+B*VK(K,*,*);
 END;
END NBPDMA;
```

Procedure ECNBPG - Energy Conserving Solution of the N-Body Problem Based on the Gragg Method

1. Application

The procedure ECNBPG integrates the equations of motion (2.1.1)-(2.1.2) of N material points in the inertial frame of reference in the interval $[t_k,t_{k+1}]$ (k=0,1,...) by the polynomial extrapolation method based on the Gragg method. The energy conserving modification is applied.

2. Data

P - one half of desired order of the method;
G - gravitational constant;
TK, TK1 - ends of the integration interval $[t_k,t_{k+1}]$;
N - number of material points;
M(1:N) - masses of material points;
X(1:3,1:N),V(1:3,1:N) - coordinates and components of velocities of N material points at the moment t_k;
EPS - accuracy in the iteration process for solving the equations (2.5.6).

3. Results

X(1:3,1:N),V(1:3,1:N) - coordinates and components of velocities of N material points at the moment t_{k+1}.

4. Procedure

```
ECNBPG: PROC (P,G,TK,TK1,N,M,X,V,EPS);
 DCL (BETA(O:P) CTL,BETA1,P,P1,Q) BIN FIXED,
     (A,AP,A1,B,C,DF,EPS,EO,FF,G,H,M(*),R,R1,R2,SUM,SUM1,TK,
      TK1,V(*,*),X(*,*),(ALFA1(I),ALFA2(I),BK(I),F(P),F1(P),
      U(3,N),VK(I,3,N),V1(3,N),W(3,N),XK(I,3,N),X1(3,N),
      Y(3,N),Z(3,N)) CTL) FLOAT (16);
 I=P+1;
 ALLOC ALFA1,ALFA2,BETA,BK,F,F1,U,VK,V1,W,XK,X1,Y,Z;
 BETA(O)=1; BETA(1)=2; BETA(2)=3; EO=0;
 DO K=3 TO P;
  BETA(K)=2 BETA(K-2);
 END;
```

```
DO I=1 TO N;
 SUM=0;
 DO J=1 TO I-1,I+1 TO N;
  SUM=SUM+M(J)/SQRT((X(1,I)-X(1,J))**2+(X(2,I)-X(2,J))**2
      +(X(3,I)-X(3,J))**2);
 END;
 EO=EO+M(I)*( V(1,I)**2+V(2,I)**2+V(3,I)**2-G*SUM);
END;
EO=EO/2; BK=1;
DO K=1 TO P+1; DO L=1 TO K-1,K+1 TO P+1;
 BK(K)=BK(K)*BETA(L-1)**2;
END; END;
F,F1=1;
DO L=1 TO P-1;
 DO I=1 TO L-1,L+1 TO P-1;
  A=BK(L)-BK(I); F(L)=F(L)*(BK(P)-BK(I))/A;
  F1(L)=F1(L)*(BK(P+1)-BK(I))/A;
 END;
 F(L)=BK(P)*F(L)/BK(L); F1(L)=BK(P+1)*F1(L)/BK(L);
END;
A,C=0;
DO L=1 TO P-1;
 A=A+F1(L); C=C+F(L);
END;
A=1-A; C=1-C; ALFA1(P+1)=0; ALFA1(P)=1/C; ALFA2(P+1)=1;
ALFA2(P)=-A/C;
DO L=1 TO P-1;
 ALFA1(L)=-F(L)/C; ALFA2(L)=A*F(L)/C-F1(L);
END;
SUM,SUM1=0;
DO K=1 TO P+1;
 A=BETA(K-1)**( 2*P); SUM=SUM+ALFA1(K)/A;
 SUM1=SUM1+ALFA2(K)/A;
END;
A1=-SUM/SUM1;
DO K=1 TO P+1;
 Z=X; U=V; BETA1=2*BETA(K-1); Y=Z+H*U/BETA1;
 DO L=1,2,3; DO I=1 TO N;
  SUM=0;
  DO J=1 TO I-1,I+1 TO N;
   R=0;
   DO Q=1,2,3;
    R=R+(Z(Q,I)-Z(Q,J))**2;
   END;
   SUM=SUM+M(J)*(Z(L,I)-Z(L,J))/R**1.5;
  END;
  W(L,I)=U(L,I)-H*G*SUM/BETA1;
 END; END;
 DO P1=2 TO BETA1-1;
  X1=Y; V1=W; Y=Z+2*H*V1/BETA1;
  DO L=1,2,3; DO I=1 TO N;
   SUM=0;
```

```
    DO J=1 TO I-1,I+1 TO N;
     R=0;
     DO Q=1,2,3;
      R=R+(X1(Q,I)-X1(Q,J))**2;
     END;
     SUM=SUM+M(J)*(X1(L,I)-X1(L,J))/R**1.5;
    END;
    W(L,I)=U(L,I)-2*H*G*SUM/BETA1;
   END; END;
   Z=X1; U=V1;
  END;
  DO L=1,2,3; DO I=1 TO N;
   SUM,SUM1=0;
   DO J=1 TO I-1,I+1 TO N;
    R,R1=0;
    DO Q=1,2,3;
     R=R+(Y(Q,I)-Y(Q,J))**2;
     R1=R1+(Z(Q,I)-Z(Q,J)+2*H*(W(Q,I)-W(Q,J))/BETA1))**2;
    END;
    SUM=SUM+M(J)*(Y(L,I)-Y(L,J))/R**1.5;
    SUM1=SUM1+M(J)*(Z(L,I)-Z(L,J)+2*H*(W(L,I)-W(L,J))/BETA1)
         /(2*R1**1.5);
   END;
   SUM1=SUM1+SUM;
   X1(L,I)=(Y(L,I)+Z(L,I))/2+H*(W(L,I)+U(L,I)/2-H*G*SUM/BETA1)
          /BETA1;
   V1(L,I)=(W(L,I)+U(L,I))/2-H*G*SUM1/BETA1;
  END; END;
  XK(K,*,*)=X1; VK(K,*,*)=V1;
 END;
 AP=A1;
LAB: FF,DF=0;
 DO I=1 TO N;
  SUM,SUM1=0;
  DO J=1 TO I-1,I+1 TO N;
   R,C=0;
   DO L=1,2,3;
    R1,R2=0;
    DO K=1 TO P+1;
     B=XK(K,L,I)-XK(K,L,J); R1=R1+(ALFA1(K)+ALFA2(K)*A1)*B;
     R2=R2+ALFA2(K)*B;
    END;
    R=R+R1**2; C=C+R1*R2;
   END;
   SUM=SUM+M(J)/SQRT(R); SUM1=SUM1+M(J)*C/R**1.5;
  END;
  R,C=0;
  DO L=1,2,3;
   R1,R2=0;
   DO K=1 TO P+1;
    B =VK(K,L,I); R1=R1+(ALFA1(K)+ALFA2(K)*A1)*B;
    R2=R2+ALFA2(K)*B;
   END;
```

```
   R=R+R1**2; C=C+R1*R2;
  END;
  FF=FF+M(I)*(R-G SUM); DF=DF+M(I)*( 2*C+G*SUM1);
 END;
 A1=A1-(FF-2*EO)/DF;
 IF ABS(A1-AP)>=EPS THEN DO;
                          AP=A1; GO TO LAB;
                         END;
 X,V=0;
 DO K=1 TO P+1;
  B=ALFA1(K)+ALFA2(K)*A1; X=X+B*XK(K,*,*); V=V+B*VK(K,*,*);
 END;
END ECNBPG;
```

Procedure RMNBDM - Relative Motion of N Bodies; Solution by the Discrete Mechanics Method

1. Application

The procedure RMNBDM integrates the equations of relative motion (3.1.4) of N-1 material points with respect to the N-th point in the time interval $[t_k, t_{k+1}]$ (k=0,1,...) by the discrete mechanics method (2.3.4)-(2.3.5),(3.3.1) with an automatic step size correction.

2. Data

G - gravitational constant;
TK,TK1 - ends of the integration interval $[t_k, t_{k+1}]$;
H - initial integration step size in the interval $[t_k, t_{k+1}]$ (for k=0 it can be accepted $H=t_1-t_o$);
N - number of materials poi
M(1:N) - masses of material points;
X(1:3,1:N-1),V(1:3,1:N-1) - coordinates and components of velocities of N-1 material points at the moment t_k;
EPS - admissible discretization error;
EPS1 - accuracy in the iteration process.

3. Results

H - optimal step size for begining of integration in the next interval;
X(1:3,1:N-1),V(1:3,1:N-1) - coordinates and components of velocities of N-1 material points at the moment t_{k+1}.

4. Procedure

```
RMNBDM: PROC (G,TK,TK1,H,N,M,X,V,EPS,EPS1);
 DCL (A,B,EPS,EPS1,G,H,HV,H1,M(*),R,RN,RN1,R1,T,TK,TK1,
      V(*,*),X(*,*),(U(3,N1),UP(3,N1),VP(3,N1),V1(3,N1),
      W(3,N1),XP(3,N1),X1(3,N1),Y(3,N1),Z(3,N1),ZP(3,N1)
      CTL) FLOAT (16),
     (S.F) BIT (1),
     P BIN FIXED;
 N1=N-1;
```

```
 ALLOC U,UP,VP,V1,W,XP,X1,Y,Z,ZP;
 S='1'B; T=TK;
LAB1: XP,X1=X; VP=V;
LAB2: DO L=1,2,3; DO I=1 TO N1;
       A=X1(L,I)+X(L,I); SUM,SUM1=0;
       DO J=1 TO I-1,I+1 TO N1;
        B=X1(L,J)+X(L,J); R,R1,RN,RN1=0;
        DO P=1,2,3;
         R=R+(X(P,I)-X(P,J))**2; R1=R1+(X1(P,I)-X1(P,J))**2;
         RN=RN+X(P,J)**2; RN1=RN1+X1(P,J)**2;
        END;
        R=SQRT(R); R1=SQRT(R1); RN=SQRT(RN); RN1=SQRT(RN1);
        SUM=SUM+M(J)*((A-B)/(R*R1*(R+R1))+B/(RN*RN1*(RN
            +RN1)));
        SUM1=SUM1+M(J)*(1-(A-B)*(X1(L,I)-X1(L,J))*(1/R1
             +1/(R+R1))/R1)/(R*R1*(R+R1));
       END;
       R,R1=0;
       DO P=1,2,3;
        R=R+X(P,I)**2; R1=R1+X1(P,I)**2;
       END;
       R=SQRT(R); R1=SQRT(R1);
       X1(L,I)=X1(L,I)-(X1(L,I)-X(L,I)-H*(V(L,I)-H*G
               *((M(N)+M(I))*A/(R*R1*(R+R1))+SUM)/2)
               /(1+H*H*G*((M(N)+M(I))*(1-A*X1(L,I)
               *(1/R1+1/(R+R1))/R1)/(R*R1*(R+R1)))/2);
      END; END;
 V1=2*(X1-X)/H-V;
 DO L=1,2,3; DO I=1 TO N1;
  IF ABS(X1(L,I)-XP(L,I))>=EPS1 | ABS(V1(L,I)-VP(L,I))>=EPS1
   THEN DO;
         XP=X1; VP=V1; GO TO LAB2;
        END;
 END; END;
 IF S='0'B THEN GO TO LAB5;
 F='1'B; Y=X; W=V;
LAB3: ZP,Z=Y; UP=W;
LAB4: DO L=1,2,3; DO I=1 TO N1;
       A=Z(L,I)+Y(L,I); SUM,SUM1=0;
       DO J=1 TO I-1,I+1 TO N1;
        B=Z(L,J)+Y(L,J); R,R1,RN,RN1=0;
        DO P=1,2,3;
         R=R+(Y(P,I)-Y(P,J))**2; R1=R1+(Z(P,I)-Z(P,J))**2;
         RN=RN+Y(P,J)**2; RN1=RN1+Z(P,J)**2;
        END;
        R=SQRT(R); R1=SQRT(R1); RN=SQRT(RN); RN1=SQRT(RN1);
        SUM=SUM+M(J)*((A-B)/(R*R1*(R+R1))+B/(RN*RN1
            *(RN+RN1)));
        SUM1=SUM1+M(J)*(1-(A-B)*(Z(L,I)-Z(L,J))*(1/R1
             +1/(R+R1))/R1)/(R*R1*(R+R1));
       END;
       R,R1=0;
```

```
      DO P=1,2,3;
       R=R+Y(P,I)**2; R1=R1+Z(P,I)**2;
      END;
      R=SQRT(R); R1=SQRT(R1);
      Z(L,I)=Z(L,I)-(Z(L,I)-Y(L,I)-H*(W(L,I)-H*G*(((M(N)
            +M(I))*A/(R*R1*(R+R1))+SUM)/4)/2)/(1+H*H*G
            *((M(N)+M(I))*(1-A*Z(L,I)*(1/R1+1/(R+R1))
            /R1)/(R*R1*(R+R1)))/2);
     END; END;
 U=4*(Z-Y)/H-W;
 DO L=1,2,3; DO I=1 TO N1;
  IF ABS(Z(L,I)-ZP(L,I))>=EPS1 | ABS(U(L,I)-UP(L,I))>=EPS1
   THEN DO;
        ZP=Z; UP=U; GO TO LAB4;
       END;
 END; END;
 IF F THEN DO;
            F='0'B; Y=Z; W=U; GO TO LAB3;
           END;
 A=0;
 DO L=1,2,3; DO I=1 TO N1;
  A=MAX(A,ABS(X1(L,I)-Z(L,I)),ABS(V1(L,I)-U(L,I)));
 END; END;
 HV=H/SQRT(2*A/EPS);
 IF HV<=H/3 THEN DO;
                  H=2*HV; GO TO LAB1;
                 END;
 IF HV>=H THEN DO;
                IF T+2*H<=TK1 THEN DO;
                                    H=2*H; GO TO LAB1;
                                   END;
                H1=H; H=TK1-T; S='0'B; GO TO LAB1;
               END;
 X=Z; V=U; T=T+H;
 IF T+2*HV<TK1 THEN DO;
                     H=2*HV; GO TO LAB1;
                    END;
 H=TK1-T; H1=2*HV; S='0'B; GO TO LAB1;
LAB5: H=H1; X=X1; V=V1;
END RMNBDM;
```

Procedure MEQPDM - Solution of the Equations of Motion nearby the Equilibrium Points by the Discrete Mechanics Method

1. Application

The procedure MEQPDM solves the equations of motion (4.2.2) of a body with infinitely small mass nearby the equilibrium point L_j (j=1,2,3,4,5) by the discrete mechanics method (4.4.3)-(4.4.4). The coordinates a_{1j}, a_{2j} and a_{3j} of the point L_j can be calculated by the following procedure:

```
COEQP: PROC (M1,M2,J,EPS,A);
 DCL (A(3),AL(0:5),A1,EPS,G,M1,M2,MI) FLOAT (16);
 G=1/(M1+M2); MI=M2*G; A(3)=0;
 IF J>3 THEN DO;
                 A(1)=G*(M1-M2)/2; A(2)=SQRT(3)/2;
                 IF J=5 THEN A(2)=-A(2); GO TO FIN;
                END;
 A(2)=0; AL(3)=(6*MI-6)*MI+1; AL(4)=4*MI-2;
 IF J=1 THEN DO;
                 AL(0)=((2*MI-3)*MI+3)*MI-1;
                 AL(1)=(((MI-2)*MI+5)*MI-4)*MI+2;
                 AL(2)=((4*MI-6)*MI+4)*MI-1;
                END;
 IF J=2 THEN DO;
                 AL(0)=(-3*MI+3)*MI-1;
                 AL(1)=(((MI-2)*MI+1)-4)*MI+2;
                 AL(2)=((4*MI-6)*MI+2)*MI-1;
                END;
 IF J=3 THEN DO;
                 AL(0)=(3*MI-3)*MI+1;
                 AL(1)=(((MI-2)*MI+1)*MI+4)*MI-2;
                 AL(2)=((4*MI-6)*MI+2)*MI+1; A1=-1-MI;
 IF J<3 THEN A1=1-MI;
LAB: G=A1; MI=5 A1;
 DO L=4 TO 0 BY -1;
  G=(G+AL(L))*A1; MI=(MI+L*AL(L))*A1;
 END;
 A(1)=A1*(1-G/MI);
 IF ABS(A(1)-A1)<EPS THEN GO TO FIN;
 A1=A(1); GO TO LAB;
FIN: END COEQP;
```

where M1, M2 denote finite masses, J is the equilibrium point number, EPS denotes an accuracy of iteration process and the array A(1:3) contains the required coordinates. The procedure MEQPDM applies an automatic step size correction.

2. Data

M1,M2 - finite masses;
A(1:3) - coordinates of equilibrium point;
TK,TK1 - ends of the integration interval $[t_k, t_{k+1}]$;
H - initial integration step size in the interval $[t_k, t_{k+1}]$ (for k=0 it can be accepted $H=t_1-t_o$);
X(1:3),V(1:3) - coordinates and components of velocity of a body with infinitely small mass at the moment t_k;
EPS - admissible discretization error;
EPS1 - accuracy in the iteration process.

3. Results

H - optional step size for beginning of integration in the

next interval;
X(1:3),V(1:3) - coordinates and components of velocity of the body with infinitely small mass at the moment t_{k+1}.

4. Procedure

```
MEQDM: PROC (M1,M2,A,TK,TK1,H,X,V,EPS,EPS1);
 DCL (A(3),AK,ALFA,BETA,BK,CK,DA(3),DB(3),DC,DF(3),EPS,EPS1,
      F(3),G,H,HV,H1,H2,MI,M1,M2,R1,R11,R2,R21,T,TK,TK1,
      U(3),UP(3),V(3),VP(3),V1(3),W(3),X(3),XP(3),X1(3),
      Y(3),Z(3),ZP(3)) FLOAT (16),
     (S,FB) BIT (1);
 G=1/(M1+M2); MI=G*M2; S='1'B; T=TK;
LAB1: XP,X1=X; VP=V; H2=H*H; R1,R2=(X(2)+A(2))**2+X(3)**2;
 R1=SQRT((X(1)+A(1)+MI)**2+R1);
 R2=SQRT((X(1)+A(1)+MI-1)**2+R2);
LAB2: R11,R21=(X1(2)+A(2))**2+X1(3)**2;
 R11=SQRT((X1(1)+A(1)+MI)**2+R11);
 R21=SQRT((X1(1)+A(1)+MI-1)**2+R21); AK=M1/(R1*R11*(R1+R11));
 BK=M2/(R2*R21*(R2+R21)); CK=2*(MI*AK-(1-MI)*BK);
 DO L=1,2,3;
  IF L=1 THEN DO;
              ALFA=MI; BETA=MI-1;
             END;
         ELSE ALFA,BETA=0;
  DA(L)=-AK((X1(L)+A(L)+ALFA)*(1/R11+1/(R1+R11))/R11;
  DB(L)=-BK((X1(L)+A(L)+BETA)*(1/R21+1/(R2+R21))/R21;
 END;
 DC=2*(MI*DA(1)-(1-MI)*DB(1));
 DO L=1,2,3;
  ALFA=X1(L)+X(L)+2*A(L); F(L)=-G*ALFA*(AK+BK);
  DF(L)=-G*(AK+BK+ALFA*(DA(L)+DB(L)));
 END;
 F(1)=F(1)-G*CK; DF(1)=DF(1)-G*DC;
 X1(1)=X1(1)-((H2/2-2)*X1(1)+(H2/2+2)*X(1)+2*H*(V(1)+X1(2)
        -X(2))+H2*(A(1)+F(1)))/(H2/2-2+H2*DF(1));
 X1(2)=X1(2)-((H2/2-2)*X1(2)+(H2/2+2)*X(2)+2*H*(V(2)-X1(1)
        +X(1))+H2*(A(2)+F(2)))/(H2/2-2+H2*DF(2));
 X1(3)=X1(3)-(2*(X(3)-X1(3))+2*H*V(3)+H2*F(3))/(H2*DF(3)-2);
 V1=2*(X1-X)/H-V;
 DO L=1,2,3;
  IF ABS(X1(L)-XP(L))>=EPS1 | ABS(V1(L)-VP(L))>=EPS1
   THEN DO;
         XP=X1; VP=V1; GO TO LAB2;
        END;
 END;
 IF S='0'B THEN GO TO LAB5;
 FB='1'B; Y=X; W=V;
LAB3: ZP,Z=Y; UP=W; R1,R2=(Y(2)+A(2))**2+Y(3)**2;
 R1=SQRT((Y(1)+A(1)+MI)**2+R1);
 R2=SQRT((Y(1)+A(1)+MI-1)**2+R2);
LAB4: R11,R21=(Z(2)+A(2))**2+Z(3)**2;
 R11=SQRT((Z(1)+A(1)+MI)**2+R11);
```

```
R21 =SQRT((Z(1)+A(1)+MI-1)**2+R21);
AK=M1/(R1*R11*(R1+R11)); BK=M2/(R2*R21*(R2+R21));
CK=2*(MI*AK-(1-MI)*BK);
DO L=1,2,3;
 IF L=1 THEN DO;
             ALFA=MI; BETA=MI-1;
            END;
        ELSE ALFA,BETA=0;
 DA(L)=-AK*(Z(L)+A(L)+ALFA)*(1/R11+1/(R1+R11))/R11;
 DB(L)=-BK*(Z(L)+A(L)+BETA)*(1/R21+1/(R2+R21))/R21;
END;
DC=2*(MI*DA(1)-(1-MI)*DB(1));
DO L=1,2,3;
 ALFA=Z(L)+Y(L)+2*A(L); F(L)=-G*ALFA*(AK+BK);
 DF(L)=-G*(AK+BK+ALFA*(DA(L)+DB(L)));
END;
F(1)=F(1)-G*CK; DF(1)=DF(1)-G*DC;
Z(1)=Z(1)-((H2/8-2)*Z(1)+H*(W(1)+Z(2)-Y(2))+(H2/8+2)*Y(1)
     +H2*(A(1)+F(1))/4)/(H2/8-2+H2*DF(1)/4);
Z(2)=Z(2)-((H2/8-2)*Z(2)+H*(W(2)-Z(1)+Y(1))+(H2/8+2)*Y(2)
     +H2*(A(2)+F(2))/4)/(H2/8-2+H2*DF(2)/4);
Z(3)=Z(3)-(2*(Y(3)-Z(3))+H*W(3)+H2*F(3)/4)/(H2*DF(3)/4-2);
U=4*(Z-Y)/H-W;
DO L=1,2,3;
 IF ABS(Z(L)-ZP(L))>=EPS1 | ABS(U(L)-UP(L))>=EPS1
  THEN DO;
        ZP=Z; UP=U; GO TO LAB4;
       END;
END;
IF FB THEN DO;
            FB='0'B; Y=Z; W=U; GO TO LAB3;
           END;
AK=0;
DO L=1,2,3;
 AK=MAX(AK,ABS(X1(L)-Z(L)),ABS(V1(L)-U(L)));
END;
HV=H/SQRT(2*AK/EPS);
IF HV<=H/3 THEN DO;
                 H=2*HV; GO TO LAB1;
                END;
IF HV>=H THEN DO;
               IF T+2*H<=TK1 THEN DO;
                                   H=2*H; GO TO LAB1;
                                  END;
               H1=H; H=TK1-T; S='0'B; GO TO LAB1;
              END;
X=Z; V=U; T=T+H;
IF T+2*HV<TK1 THEN DO;
                    H=2*HV; GO TO LAB1;
                   END:
H1=2*HV; S='0'B; H=TK1-T;
IF H=0 THEN GO TO LAB6;
```

```
 GO TO LAB1;
LAB5: X=X1; V=V1;
LAB6: H=H1;
END MEQDM;
```

Procedure DHEQ - Solution of Discrete Hill's Equations

1. Application

The procedure DHEQ solves the discrete Hill's equations (4.5.3) in the time interval $[\tau_k, \tau_{k+1}]$ (k=0,1,...) (see (4.1.29)) with an automatic step size correction.

2. Data

G - gravitational constant;
MO - mass of the earth;
M1 - mass of the sun;
NI - mean motion of the moon;
NI1 - mean motion of the sun;
TAUK,TAUK1 - ends of the integration interval $[\tau_k, \tau_{k+1}]$;
H - initial integration step size in the interval $[\tau_k, \tau_{k+1}]$ (for k=0 it can be accepted $H=\tau_1-\tau_0$);
X(1:3),V(1:3) - coordinates and components of velocity of the moon at the moment τ_k;
EPS - admissible discretization error;
EPS1 - accuracy in the iteration process.

3. Results

H - optimal step size for beginning of integration in the next interval;
X(1:3),V(1:3) - coordinates and components of velocity of the moon at the moment τ_{k+1}.

4. Procedure

```
DHEQ: PROC (G,MO,M1,NI,NI1,TAUK,TAUK1,H,X,V,EPS,EPS1);
 DCL (A,DF(3),EPS,EPS1,F(3),G,H,HM,HV,H1,H2,KAPPA,M,MO,M1,
      NI,NI1,R,R1,TAU,TAUK,TAUK1,U(3),UP(3),V(3),VP(3),
      V1(3),W(3),X(3),XP(3),X1(3),Y(3),Z(3),ZP(3))
      FLOAT (16),
     (S,FB) BIT (1);
 KAPPA=G*(MO+M1)/(NI-NI1)**2; M=NI1/(NI-NI1); S='1'B;
 TAU=TAUK;
LAB1: XP,X1=X; VP=V; H2=H H; HM=M*M*H*H; R=0;
 DO L=1,2,3;
  R=R+X(L)*X(L);
 END;
 R=SQRT(R);
LAB2: R1=0;
 DO L=1,2,3;
  R1=R1+X1(L)*X1(L);
 END;
```

```
 DO L=1,2,3;
  F(L)=-KAPPA*(X1(L)+X(L))/(R*R1*(R+R1));
  DF(L)=KAPPA*(-1+X1(L)*(X1(L)+X(L))*(1/R1+1/(R+R1))/R1)
        /(R*R1*(R+R1));
 END;
 F(1)=F(1)+M*M*(X1(1)+X(1)); DF(1)=DF(1)+M*M;
 DO L=2,3;
  F(L)=F(L)-M*M*(X1(L)+X(L))/2; DF(L)=DF(L)-M*M/2;
 END;
 X1(1)=X1(1)-((HM/2-2)*X1(1)+2*H*(V(1)+M*(X1(2)-X(2)))
       +(HM/2+2)*X(1)+H2*F(1))/(HM/2-2+H2*DF(1));
 X1(2)=X1(2)-((HM/2-2)*X1(2)+2*H*(V(2)-M*(X1(1)-X(1)))
       +(HM/2+2)*X(2)+H2*F(2))/(HM/2-2+H2*DF(2));
 X1(3)=X1(3)-(2*(X(3)-X1(3)+H*V(3))+H2*F(3))/(H2*DF(3)-2);
 V1=2*(X1-X)/H-V;
 DO L=1,2,3;
  IF ABS(X1(L)-XP(L))>=EPS1 | ABS(V1(L)-VP(L))>=EPS1
   THEN DO;
         XP=X1; VP=V1; GO TO LAB2;
        END;
 END;
 IF S='0'B THEN GO TO LAB5;
 FB='1'B; Y=X; W=V;
LAB3: ZP,Z=Y; UP=W; R=0;
 DO L=1,2,3;
  R=R+Y(L)*Y(L);
 END;
 R=SQRT(R);
LAB4: R1=0;
 DO L=1,2,3;
  R1=R1+Z(L)*Z(L);
 END;
 R1=SQRT(R1);
 DO L=1,2,3;
  F(L)=-KAPPA*(Z(L)+Y(L))/(R*R1*(R+R1));
  DF(L)=KAPPA*(-1+Z(L)*(Z(L)+A(L))*(1/R1+1/(R+R1))/R1)
        /(R*R1*(R+R1));
 END;
 F(1)=F(1)+M*M*(Z(1)+Y(1)); DF(1)=DF(1)+M*M;
 DO L=2,3;
  F(L)=F(L)-M*M*(Z(L)+Y(L))/2; DF(L)=DF(L)-M*M/2;
 END;
 Z(1)=Z(1)-((HM/8-2)*Z(1)+H*(W(1)+M*(Z(2)-Y(2)))+(HM/8+2)*Y(1)
     +H2*F(1)/4)/(HM/8-2+H2*F(1)/4);
 Z(2)=Z(2)-((HM/8-2)*Z(2)+H*(W(2)-M*(Z(1)-Y(1)))+(HM/8+2)*Y(2)
     +H2*F(2)/4)/(HM/8-2+H2*F(2)/4);
 Z(3)=Z(3)-(2*(Y(3)-Z(3)+H*W(3))+H2*F(3)/4)/(H2*F(3)/4-2);
 U=4*(Z-Y)/H-W;
 DO L=1,2,3;
  IF ABS(Z(L)-ZP(L))>=EPS1 | ABS(U(L)-UP(L))>=EPS1
   THEN DO;
         ZP=Z; UP=U; GO TO LAB4;
        END;
```

```
END;
IF FB THEN DO;
            FB='0'B; Y=Z; W=U; GO TO LAB3;
           END;
A=0;
DO L=1,2,3;
 A=MAX(A,ABS(X1(L)-Z(L)),ABS(V1(L)-U(L)));
END;
HV=H/SQRT(2*A/EPS);
IF HV<=H/3 THEN DO;
                 H=2*HV; GO TO LAB1;
                END;
IF HV>=H THEN DO;
               IF T+2*H<=TAUK1 THEN DO;
                                     H=2*H; GO TO LAB1;
                                    END;
               H1=H; H=TAUK1-TAU; S='0'B; GO TO LAB1;
              END;
X=Z; V=U; TAU=TAU+H;
IF TAU+2*HV<TAU1 THEN DO;
                       H=2*HV; GO TO LAB1;
                      END;
H1=2*HV; S='0'B; H=TAUK1-TAU;
IF H=0 THEN GO TO LAB6;
GO TO LAB1;
LAB5: X=X1; V=V1;
LAB6: H=H1;
END DHEQ;
```

References

[1] Albrycht,J.,Marciniak,A., Orbit calculations nearby the equilibrium points by a discrete mechanics method, Cel. Mech.24 (1981), 391-405.

[2] Albrycht,J.,Marciniak,A., Discrete dynamical equations in Minkowski space, Int.J.Theor.Physics 20 (1981), 821-830.

[3] Albrycht,J.,Marciniak,A., Asymptotic expansion of total error for a discrete mechanics method (unpublished).

[4] Bahvalov,N.S., Numerical Methods [in Russian], Izdatel'stvo "Nauka", Moscow 1973.

[5] Barton,D.,Willers,I.M.,Zahar,R.U.M., Taylor series methods for ordinary differential equations - an evaluation, in: Mathematical Software (Rice,J.R.,ed.), Academic Press, New York 1971.

[6] Belorizky,D., Sur la convergence des séries dans la solution du problème des trois corps donée per M.Sundman, Comptes Rendus Hebdomaires de l'Acad. des Sci.193 (1931), 314 et seqq.

[7] Belorizky,D., Application pratique des méthodes de M.Sundman à un cas particulier des trois corps, Bull. Astronomique 6 (1931), 417 et seqq.

[8] Belorizky,D., Recherches sur l'application pratique des solutions générales du problème des trois corps, J.Observateurs 16 (1933), 109 et seqq.

[9] Brouwer,D.,Clemence,G.M., Methods of Celestial Mechanics, Academic Press, New York 1961.

[10] Bulirsch,R.,Stoer,J., Fehlerabschatzungen und Extrapolation mit rationalen Funktionen bei Verfahren vom Richardson-Typus, Num.Math.6 (1964), 413-427.

[11] Bulirsch,R.,Stoer,J., Numerical treatment of ordinary differential equations by extrapolation methods, Num. Math.8 (1966), 1-13.

[12] Butcher,J.C., Implicit Runge-Kutta processes, Math.Comp. 18 (1964), 50-64.

[13] Butcher,J.C., On Runge-Kutta processes of high order,

J.Austral.Math.Soc.4 (1964), 179-194.
[14] Butcher,J.C., A modified multistep method for the numerical integration of ordinary differential equations, J.Ass.Comput.Mach.12 (1965), 124-135.
[15] Butcher,J.C., The effective order of Runge-Kutta methods, Conf.on the numer.sol.of diff.eqns., Lecture Notes in Math.109, 133-139, Springer, Berlin 1969.
[16] Coddington,A.,Levinson,N., Theory of Ordinary Differential Equations, McGraw-Hill, New York 1965.
[17] Contopoulos,G., A third integral of motion in a galaxy, Z.Astrophys.49 (1960), 273 et seqq.
[18] Contopoulos,G., On the existence of a third integral of motion, Astronomical J.68 (1963), 1 et seqq.
[19] Crane,R.L., Klopfenstein,R.W., A predictor-corrector algorithm with an increased range of absolute stability, J.Ass.Comput.Mach.12 (1965), 227-241.
[20] Curtis,A.R., An eight order Runge-Kutta process with eleven function evaluation per step, Num.Math.16 (1970), 268-277.
[21] Dahlquist,G.,Björck,A., Numerical Methods, Prentice Hall, New Jersey 1976.
[22] Dubošin,G.N., Celestial Mechanics. Basic Problems and Methods [in Russian], Izdatel'stvo "Nauka", Moscow 1963.
[23] Dubošin,G.N., Celestial Mechanics. Analitical and Computational Methods [in Russian], Izdatel'stvo "Nauka", Moscow 1964.
[24] England,R., Error estimates for Runge-Kutta type solutions to systems of O.D.E's., Comput.J.12 (1969), 166-169.
[25] Fehlberg,E., New high-order Runge-Kutta formulas with stepsize control for systems of first and second order differential equations, Z.Angew.Math.Mech.44 (1964), T17-T29.
[26] Fehlberg,E., New high-order Runge-Kutta formulas with an arbitrary small truncation error, Z.Angew.Math.Mech. 46 (1966), 1-16.
[27] Fehlberg,E., Classical fifth, sixth, seventh and eight order Runge-Kutta formulas with stepsize control, NASA TR 287, 1968.
[28] Finley-Freundlich,E., Celestial Mechanics, London 1958.
[29] Fox,P.A., DESUB: Integration of a first-order system of ordinary differential equations, in: Math.Software (Rice,J.R.,ed.), Academic Press, New York 1971.
[30] Gear,C.W., The numerical integration of ordinary differential equations, Math.Comp.21 (1967), 146-156.
[31] Gear,C.W., The automatic integration of ordinary differential equations, Comm.Ass.Comput.Mach.14 (1971), 176-179.
[32] Gear,C.W., Numerical Initial Value Problems in Ordinary Differential Equations, Prentice-Hall, New Jersey 1971.
[33] Gear,C.W.,Tu,K.W., The effect of variable mesh size on the stability of multistep methods, SIAM J.Numer.

Anal.11 (1974), 1025-1043.
[34] Gragg,W.B.,Stetter,H.J., Generalized multistep predictor-corrector methods, J.Ass.Comput.Mach.11 (1964), 189-209.
[35] Gragg,W.B., On extrapolation algorithms for ordinary initial value problems, SIAM J.Numer.Anal.2 (1965), 384-403.
[36] Greenspan,D., New forms of discrete mechanics, Kybernetes 1 (1972), 87-101.
[37] Greenspan,D., A new explicit discrete mechanics with applications, J.Franklin Inst.294 (1972), 231-240.
[38] Greenspan,D., Discrete Newtonian gravitation and the N-body problem, Utilitas Math.2 (1972), 105-126.
[39] Greenspan,D., An algebraic, energy conserving formulation of classical molecular and Newtonian N-body interaction, Bull.Amer.Math.Soc.2 (1973), 423-427.
[40] Greenspan,D., Symmetry in discrete mechanics, Found. Physics 2 (1973), 247-253.
[41] Greenspan,D., Discrete Numerical Methods in Physics and Engineering, Academic Press, New York 1974.
[42] Greenspan,D., Arithmetic Applied Mathematics, Pergamon Press, Oxford 1980.
[43] Hall,G., Stability analysis of predictor-corrector algorithms of Adams type, SIAM J.Numer.Anal.11 (1974), 494-505.
[44] Hall,G.,Watt,J.M.,eds., Modern Numerical Methods for Ordinary Differential Equations, Clarendron Press, Oxford 1976.
[45] Hall,G.,et al., DETEST: a program for comparing numerical methods for ordinary differential equations, Univ.of Toronto, Dept.of Comput.Sci., TR No.60.
[46] Hamming,R.W., Stable predictor-corrector methods for ordinary differential equations, J.Ass.Comput.Mach. 6 (1959), 37-47.
[47] Henrici,P., Discrete Variable Methods in Ordinary Differential Equations, J.Wiley & Sons, New York 1964.
[48] Hill,G.W., Researches in the lunar theory, Am.J.Math. 1 (1878).
[49] Hull,T.E.,Creemer,A.L., Efficiency of predictor -corrector schemes, J.Ass.Comput.Mach.10 (1963), 291-301.
[50] Hull,T.E.,Luxemberg,W.A.J., Numerical methods and existence theorems for ordinary differential equations, Num.Math.2 (1960), 30-41.
[51] Hull,T.E.,et al., Comparing numerical methods for ordinary differential equations, SIAM J.Numer.Anal.9 (1972), 603-637.
[52] Isaacson,E.,Keller,H.B., Analysis of Numerical Methods, J.Wiley & Sons, New York 1966.
[53] Jacobi,C.G.J., Sur l'élimination des noeuds dans le problème des trois corps, Comptes Rendus hebdomadaires de l'Acad.des Sci.15 (1842), 236 et seqq.

[54] Jankowska,J.,Jankowski,M., Survey of Numerical Methods and Algorithms. Part 1 [in Polish], WNT, Warsaw 1981.
[55] Kohfeld,J.J.,Thompson,G.T., Multistep methods with modified predictors and correctors, J.Ass.Comput.Mach. 14 (1967), 155-166.
[56] Krogh,F.T., Predictor-corrector methods of high order with improved stability characteristics, J.Ass.Comput. Mach.13 (1966), 374-385.
[57] Krogh,F.T., Algorithms for changing the stepsize, SIAM J.Numer.Anal.10 (1973), 949-965.
[58] Lambert,J.D.,Shaw,B., A generalization of multistep method for ordinary differential equations, Num.Math.8 (1966), 250-263.
[59] Lambert,J.D., Computational Methods in Ordinary Differential Equations, J.Wiley & Sons, London 1973.
[60] Lapidus,L.,Seinfeld,J.H., Numerical Solution of Ordinary Differential Equations, Academic Press, New York 1971.
[61] Luther,H.A.,Sierra,H.G., On the optimal choice of fourth-order Runge-Kutta formulas, Num.Math.15 (1970), 354-358.
[62] Marciniak,A., Discrete Hill's equations, Comp.Meth. Appl.Mech.Eng.37 (1983), 15-24.
[63] Marciniak,A., Discrete mechanics of arbitrary order, Comp.Meth.Appl.Mech.Eng.39 (1983), 159-178.
[64] Marciniak,A., Energy conserving, arbitrary order numerical solutions of the N-body problem, Num.Math.45 (1984), 207-218.
[65] Marciniak,A., Discrete mechanics and its application to the solution of the N-body problem, Acta Appl.Math. 2 (1984), 185-207.
[66] Marciniak,A., Application of discrete mechanics for solving the N-body problem [in Polish], Ph.D.Thesis, A.Mickiewicz Univ., Poznań, Poland.
[67] Marčuk,G.J.,Šajdurov,V.V., An Increase of Precision for Solutions of Difference Schemes [in Russian], Izdatel'stvo "Nauka", Moscow 1979.
[68] Milne,W.E., Numerical Solution of Differential Equations, J.Wiley & Sons, New York 1953.
[69] Moszyński,K., Methods of Solving Ordinary Differential Equations on a Digital Computer [in Polish], WNT, Warsaw 1971.
[70] Nordsieck,A., On numerical integration of ordinary differential equations, Math.Comp.16 (1962), 22-49.
[71] Petrovskij,I., Lectures on Ordinary Differential Equations [in Russian], 6th ed., Izdatel'stvo "Nauka", Moscow 1970.
[72] Pollard,H., Mathematical Introduction to Celestial Mechanics, Prentice Hall, New Jersey 1966.
[73] Pollard,H., Celestial Mechanics, The Math.Ass.of America, 1976.
[74] Pontriagin,L.S., Ordinary Differential Equations [in

Russian], 3rd ed., Izdatel'stvo "Nauka", Moscow 1970.
[75] Potter,D., Computational Physics, J.Wiley & Sons, London 1973.
[76] Ralston,A.,Rabinowitz,P., A First Course in Numerical Analysis, 2nd ed., McGraw-Hill, New York 1978.
[77] Rosser,J.B., A Runge-Kutta for all seasons, SIAM Rev.9 (1967), 417-452.
[78] Šarl'e,K., Celestial Mechanics [in Russian], Izdatel'stvo "Nauka", Moscow 1966.
[79] Shampine,L.F., Local extrapolation in the solution of ordinary differential equations, Math.Comp.27 (1973), 91-97.
[80] Shampine,L.F.,Gordon,M.K., Computer Solution of Ordinary Differential Equations. The Initial Value Problem, W.H.Freeman & Co., 1975.
[81] Smart,W.M., Celestial Mechanics, London 1953.
[82] Stepanov,V., A Course of Differential Equations [in Russian], 8th ed., Fizmatgiz, Moscow 1959.
[83] Sterne,T.E., An Introduction to Celestial Mechanics, Interscience, New York 1960.
[84] Stetter,H.J., Richardson-extrapolation and optimal estimation, Appl.Math.13 (1968), 187-190.
[85] Stetter,H.J., Improved absolute stability of predictor -corrector schemes, Computing 3 (1968), 286-296.
[86] Stetter,H.J., Analysis of Discretization Methods of Ordinary Differential Equations, Springer-Verlag, Berlin 1973.
[87] Stoer,J.,Bulirsch,R., Einführung in die Numerische Mathematik II, Springer-Verlag, Berlin 1973.
[88] Stoer,J., Extrapolation methods for the solution of initial value problems and their practical realization, Lecture Notes in Math.362, 1-21, Springer, New York 1974.
[89] Szebehely,V., Theory of Orbits. The Restricted Problem of Three Bodies, New York 1967.
[90] Sundman,K.F., Mémoire sur le problème des trois corps, Acta Math.36 (1912), 105 et seqq.
[91] Wierzbiński,S., Celestial Mechanics [in Polish], PWN, Warsaw 1973.

Index